MW00855941

The
Entropy
Crisis

The Entropy Crisis

Guy Deutscher

Tel Aviv University, Israel

World Scientific

NEW JERSEY • LONDON • SINGAPORE • BEIJING • SHANGHAI • HONG KONG • TAIPEI • CHENNAI

Published by

World Scientific Publishing Co. Pte. Ltd.
5 Toh Tuck Link, Singapore 596224
USA office: 27 Warren Street, Suite 401-402, Hackensack, NJ 07601
UK office: 57 Shelton Street, Covent Garden, London WC2H 9HE

British Library Cataloguing-in-Publication Data
A catalogue record for this book is available from the British Library.

Cover Illustration: Aline Deutscher

THE ENTROPY CRISIS

Copyright © 2008 by World Scientific Publishing Co. Pte. Ltd.

All rights reserved. This book, or parts thereof, may not be reproduced in any form or by any means, electronic or mechanical, including photocopying, recording or any information storage and retrieval system now known or to be invented, without written permission from the Publisher.

For photocopying of material in this volume, please pay a copying fee through the Copyright Clearance Center, Inc., 222 Rosewood Drive, Danvers, MA 01923, USA. In this case permission to photocopy is not required from the publisher.

ISBN-13 978-981-277-968-7
ISBN-10 981-277-968-X
ISBN-13 978-981-277-969-4 (pbk)
ISBN-10 981-277-969-8 (pbk)

Printed in Singapore.

To

Barak, Eyal, Nimrod, Roï, Noga and Ori

as a Vade Mecum for life in a changing world

Contents

Introduction

The biosphere is the world in which we live. Compared to the size of the earth, the biosphere is a thin layer surrounding the earth's surface, extending a few kilometers above and below it. This is where all living organisms and their residues are to be found. This is where mankind has developed, in intimate relation with its surroundings, what we often call the environment. We know that our life has always been dependent on it, and always will. Hence our deep concern when changes, maybe harmful to us, occur in the biosphere, possibly due to the activity of mankind.

Do these fears have a scientific basis, or are they grossly exaggerated? For instance, is climate change a real threat? And what is more harmful to the biosphere, to burn more fossil fuels, or to build and operate more nuclear reactors? Today these issues are in the public domain and, in the end, it will be the people who will decide what should or should not be done. This is why I believe it is important for everybody to understand the nature of the issues at hand.

It turns out that an intelligent discussion requires some familiarity with a concept called entropy. While everybody is familiar with the concept of energy, only a few, mostly scientists, know about entropy. In order to understand that there is a deep connection between the energy crisis, by which we mean that we may soon run out of fossil fuels, and damage to the environment, it is necessary to understand the concept of entropy. According to the laws of thermodynamics, this damage is one aspect of an increase in entropy (or disorder at the molecular scale) in the biosphere, which cannot be avoided when we burn fuel. This increase in entropy is more subtle than the loss of fuel supply, but instinctively it is the one that we fear more. As we shall see, the danger lies not so much in the fact that we are burning fuel, but rather in the rate at which we do this.

If we run out of fossil fuels it is evidently because we have been burning them so fast, and so inefficiently, and this is precisely the reason why the effects of the increase in entropy are now there for everybody to see.

I wish to thank Rafael Ben Zeev for prompting me to write this little book; my wife Aline for her continuing encouragements and well taken criticisms of the whole manuscript; my son Daniel for his very critical reading of Chapter 4 where the thermodynamic concept of entropy is introduced, and insisting that it should be made accessible to high school level students; to my daughter Nathalie for making some good points on the first chapters; to my friend Gila for her remarks on my way of writing English; to my friend Michael for his sharp criticism of Chapter 5; and to Jacques Friedel for several and enlightening discussions on the issues raised by the use of nuclear energy for electricity production.

Guy Deutscher
June 2008

Chapter 1

Dealing with Entropy on a Daily Basis

In elementary and even high school education, a great deal of time is spent teaching students the basic notions of force, energy and power, but the word "entropy" is often not even pronounced. It is only those who specialize in the sciences who will become familiar with it. This is a dramatic shortcoming of our education system, as without some understanding of what entropy means it is essentially impossible to comprehend what is going on in the environment and to make the right decisions for its defense.

Historically, the notion of entropy was developed to understand what happens when one transforms heat into mechanical energy. This is what one did in the steam engine: one burned coal and the engine produced work. But it appeared that only a fraction of the energy stored in the fuel being burnt was transformed into mechanical energy. Where did the rest go? On the other hand, mechanical work can be transformed into heat without any loss. Why is this? Before we deal with these questions in a further chapter, it may be useful to show by a few examples taken from our daily life that, just as Monsieur Jourdain in the play by Moliere "Le bourgeois gentilhomme" was speaking in prose without knowing it, we are every day dealing with entropy without knowing that this is what we are doing.

1.1. Entropy in the household

Physicists like to do what they call a "gedanken experiment", which consists in imagining doing a certain experiment and drawing consequences from its (virtual) outcome.

So let us do the following gedanken experiment. Imagine that in our home we stop doing all the things that we do everyday to keep it in good order. We do not make our beds anymore; do not give the chidren their bath; we do not clean the dishes; we use up all our clean sheets, underwear, shirts, dresses and so on and just pile them up, or rather let them lay anywhere in the house when they are dirty; we never use a broom, vacuum cleaner or other instruments developed to keep the house clean of dust and dirt; we never put back a paper that we have taken out of a file; never get rid of the garbage; stop paying our electricity, gas and water bills; never paint the house, inside or outside; never do any repair work; and so on. Sounds like paradise, doesn't it? But how long do you think you could hold on, before running out of the house and looking for shelter somewhere else?

Of course, remember, this is a gedanken experiment: I am not really suggesting that you do it to find out what the outcome is. This is precisely the beauty of a good gedanken experiment: you do not have to do it to see the consequences.

So what have we learned? Something we knew of course all the time, namely that our daily life can only be sustained if we keep putting things back in order. Not that we like doing it, but one quickly learns that there simply is no other way. The reason we do not like doing all these things is also pretty obvious: they all require work, or money that we usually earn by doing work (although I am aware that this may not be true for everybody, but those people who do not have to work also have servants that do all these things for them, so they are not concerned by our gedanken experiment anyway).

Restoring order, or reducing the amount of disorder, requires work. Physicists say that an energy input is necessary to lower the entropy of the system. Disorder in a household is not a quantity that we can define exactly, although we well understand what it means. Physicists have of course an exact definition for entropy, which the interested reader can read about in a later chapter.

Before giving more scientific examples of how an energy input is necessary to fight disorder, let us however continue a little bit with our example and illustrate the meaning of an entropy crisis.

1.2. An example of an entropy crisis at home

So far, our gedanken experiment was implicitly assuming that we were stopping doing all these things from our own will, simply because we did not like doing them. But what if we really *could not* do them?

Many of us have experienced how minor problems, such as our cleaning person suddenly not being available, or the momentary absence from home of the mistress of the house, can disturb our daily life. Usually things do not get out of hand, but a sense of impending crisis may be felt.

Things can be more serious. We may suffer from a long illness; we may lose our job. We may indeed not be able to pay our bills. Such situations are not uncommon, and are well known to social workers. If the mother is so depressed, or tired because she works night shifts, that she does not get up in the morning, children do not get their breakfast, nor the sandwich they are supposed to take to school; soon they will stop doing their homework, and get into trouble in many ways. If society is unable or unwilling to extend the necessary help to this family, it may eventually disintegrate with children becoming delinquent and the parents (often a single one) homeless on the street. These things do happen. Entropy has won, because insufficient resources were available from the outside world.

1.3. Where does all the disorder go?

Let us go back to our well ordered household, where the parent in charge does get up every morning, does the needful, everybody is fed, and the house is kept clean. Water, electricity and gas flow freely since we pay our bills, all appliances such as dishwasher, washing-machine and drier are operating, floors are vacuum-cleaned, life is beautiful. But where did all the disorder go?

Well in fact we know that too. Garbage was collected, used waters containing the dirt that came out from dirty dishes and laundry and chemical used for those purpose were evacuated. In short, disorder was transferred from our home to the outside world. And the outside world — society — will have to deal with it in some way. Garbage and used water will increase what we call pollution of the environment, unless more energy resources are spent to recycle or treat them. But in many locations on earth, garbage is not collected, and used water not treated because the necessary resources (in the end energy) are insufficient. Streams get contaminated, water is below drinking quality, dangerous diseases are rampant.

Control of disorder and use of energy are two aspects of the same problem. The less disorder we produce, the less energy we shall need to put things back in order. In Japan, tradition requires that shoes be left at the door step: dirt left out of the house will not need to be cleared. Likewise in my childhood in France it was customary to take your shoes off when entering an apartment and you would be given a couple of "patins" over which you would glide on the beautifully clean and shining wooden floor. If we drill a hole in a wall, it is best to collect dust immediately rather than to have to vacuum-clean the entire floor later on. A good housewife and her husband know for sure many tricks to reduce disorder to a minimum in their activities. They will also try to teach one or two things to their children so that they will not mess up too much.

The amount of disorder that can be tolerated is of course a matter of local customs and standard of living. Some housewives will not feel comfortable unless the house is cleaned from top to bottom every morning. Others will be more lenient. The amount of disorder that will be tolerated will also depend on what machines are available to help the housewife in her work, provided electricity is available and affordable to operate them, and water and the chemicals needed can be purchased at little cost. When there were no washing-machines, for instance, and the laundry had to be cleaned by hand, I imagine that underwear, shirts and so on were not declared "dirty" as quickly as they are now in western countries, and still households would operate reasonably well. So, fortunately, one does not have to be very strict as to the degree to which disorder must be controlled.

1.4. Disorder and pollution

What we have called disorder applies to all scales, from macroscopic objects down to molecules, but so far we have only mentioned specifically disorder on a scale that is visible to the naked eye, such as dirty dishes in a sink and dirty laundry. Disorder on a microscopic scale is often called pollution. We may define a polluted medium, such as air or water, as one that contains small particles and molecules, which are not supposed to be there — just as dirty dishes in the sink. Examples are easy to come by, such as too many CO_2 molecules in the atmosphere that are believed to be at the origin of global warming, or small black particles in the air of our cities that we find at the end of the day on the collars of our shirts and possibly could be seen in our lungs, if we were looking for them. Underground water reservoirs can be polluted for example by chemical products, by heavy metals, or by a high degree of salinity, due to exaggerated depletion, and become unsuitable for consumption.

We have given here examples of pollution due to human activity, but the tendency of small particles and molecules to move around and invade a medium is very general, and in a sense unbeatable. While heavy objects such as pieces of furniture will not move around by themselves and will be found exactly where we left them even years later, small particles and molecules are all the time on the go. In a ray of light entering through a window, we have all seen dust particles dancing around in an incessant ballet — up and down, left and right. Dust particles move around by themselves, or so it seems. If you use your broom to collect some in a corner of the house, chances are that if you come back a few days or weeks later you will find that many of them (the smaller ones) must have gone somewhere else because they are not where you left them. If you wait long enough, all the dust particles that you had so carefully collected, maybe with the exception of the heavier ones, will be found everywhere in the room if its door was closed, or anywhere in the apartment if doors were not closed. This is not a gedanken experiment, it really happens by itself all the time. Put the dust back in a corner, it will escape again.

1.5. Entropy and the second law of thermodynamics

We may call this the nightmare of the housewife. She is fighting a powerful genius, named entropy. This genius sees to it that if dust has been carefully collected in one corner of one room, that room and later all other rooms in the house will get dusty in the long run. Alas, nobody has ever seen dust getting collected by itself into a neat little corner. Exactly for the same reason, salt molecules in sea-water will never regroup by themselves so as to leave us with a nice fraction of desalinated water fit for our consumption. These are all aspects of the same law of nature which physicists call the second law of thermodynamics: in a closed system that does not benefit from an energy input, entropy (disorder) can only go up, never down. Energy is conserved, this is the first law of thermodynamics; but entropy is not. A corollary is that an energy input is necessary if one wants to lower entropy.

1.5.1. *Water desalination*

A good and important example is water desalination. If we add salt molecules to a container filled with clean water, they will quickly spread around in the entire container: this is the state of maximum entropy. Now suppose that the container is a cylinder in which we have placed a piston bearing a membrane that is permeable to water molecules but not to salt molecules. Water molecules can move freely through this membrane, but salt molecules cannot. Initially, the cylinder is filled with pure water, and the piston sits somewhere about the middle of the cylinder. If we add our salt molecules on one side of the piston, they will want to fill the entire cylinder. The piston will be submitted to a pressure which will move it to the other side. Mutatis mutandis, if we apply to the piston an inverse pressure we can collect back the salt molecules at one end of the cylinder. Water desalination has been achieved —

or order restored — at the cost of the work that we have done to push back the piston. Reducing entropy requires work.

1.5.2. *Heat transfer*

Another important example is that of heat transfer. Suppose you take a bottle of milk from the refrigerator, pour milk in a glass that you leave for a while on the table while returning the bottle to the refrigerator. You know from experience that if you wish to drink your milk cold, you had better drink it quickly because after some time the milk in the glass will have warmed up to the temperature of the room. After that you can wait for any amount of time you wish, the glass of milk will never get cold again. This is in fact the reason why you quickly returned the bottle of milk to the refrigerator. In a closed environment temperature will always tend to be uniform. We say that heat always flows from hot to cold regions, never the other way around.

But what exactly is heat? This is a difficult question that even Isaac Newton, probably the most remarkable scientific genius of all time, was unable to solve. He proposed that "heat particles" were doing the job, a view that should not surprise the reader since after all such particles would just behave in the same way as our salt molecules in water or dust particles in the room, spreading heat evenly in the apartment. Also, Newton liked particles, having explained how light propagates by assuming that it consisted of elementary "grains of light". In that case he was proven right, these are the photons. But in the case of heat, there are no extra particles that transfer heat through a medium. As was shown by Boltzman much later, in the second half of the nineteenth century, heat transfer in a gas is due to the motion of the molecules that constitute the gas itself. At the time of Newton no one seriously believed in the existence of molecules. This state of affairs persisted until Boltzman, and even then his theory was strongly debated.

Motion of the molecules that constitute the gas is precisely what lies behind the strange ballet of the dust particles that we saw dancing in the ray of light. Gas molecules kick small dust particles around, as we can see from their incessant ballet (although we do not see the molecules because they are too small). To go back to the mechanism of heat transfer, molecules move faster in hot regions than they do in cold ones. When fast and slow molecules collide, which they will necessarily do since they share the same space, it turns out that on the average faster molecules will transfer to the slower ones more energy than they get from them (energy of motion or kinetic energy), so that eventually the distribution of velocities will be the same everywhere and temperature will become uniform. This is the mechanism by which the system maximizes its entropy. The spontaneous flow of heat from hot to cold regions is of course very useful as this is how heat can be transformed into work and eventually into electricity. More on this in Chapter 4.

1.5.3. *Entropy and the states of matter*

Matter exists in different forms, as gas, liquid or solid phases. The highest degree of disorder is found in low density gases where molecules can occupy a very large number of different positions. Next comes the liquid phase, which is much denser and gives less "choice" to the moving molecules. Last comes the solid, where the molecules are basically fixed in space. According to this classification, the entropy counted per molecule is highest in the gas and lowest in the solid phase. When a gas cools down and transforms into a liquid — like steam transforming into water — entropy is reduced. It reduces again when the liquid phase transforms into a solid phase — like when water transforms into ice. Changes of entropy when matter is transformed from one phase to another play an important role in the global changes occurring in the biosphere.

The second law of thermodynamics is the reason why, once small particles or molecules are released in the atmosphere, they will spread, and eventually will be found anywhere in the biosphere. This is how entropy is maximized. Pollution is a global phenomenon, not a local one. This is not like our neat household that we have kept clean and in good order. We can do that even if our next door neighbor is less careful and his house is a real mess. This nice separation does not work for small particles and molecules released in the outside world, either by us or by our neighbor. We shall both have to suffer the consequences. Here the problem has gotten out of hand. There is no piston that we can push on to restore order.

1.6. From the household to the biosphere

Let us go back to our household. An energy input is necessary to keep it in good order: we take in energy from the outside world and release entropy into it. Here, the inside is the household and the outside is the biosphere. In the long run this increase in entropy in the biosphere may become problematic, as happens when garbage and used waters are not properly collected. Let us give one more example of an unhealthy increase of entropy: we need food for ourselves (energy input necessary to keep our internal entropy level from rising), which we burn with oxygen from the air, and as part of this combustion process we eject carbon di-oxide. In this process the overall entropy of the system (which comprises ourselves and the contents of the house, including stored food and oxygen) has increased. If we keep our doors and windows closed, and do not let fresh air in, we are eventually going to suffocate and we shall die although we still have stored food (energy). This is an entropy, not an energy crisis. Similarly, if the biosphere becomes too polluted life may become endangered even if we still have plenty of energy resources available.

An additional worry is that entropy is also ejected into the biosphere by most of the energy production methods used today. We do need energy to keep our house in good order, and also to keep it warm in the winter, cool in the summer, to light it up at night and so on. But methods of producing energy that increase entropy in the biosphere, such as burning fossil fuels, are dangerous and should be avoided as much as possible. We can usefully classify the various methods of energy production according to the amount of additional entropy they produce. Such a classification will be given in a later chapter.

Running out of energy supplies may indeed be a concern, but increasing the entropy in the biosphere may be an even more serious and more immediate one.

Chapter 2

A Short History of the Biosphere

The history of the biosphere is a long and complex one, as it extends over a period of time of several billion years and involves powerful astronomical and geological events, drastic changes in the atmosphere as well as the evolution of species. The Biblical chapter Genesis gives us the shortest account of this history in an optimistic perspective that emphasizes all the beautiful things that surround us: the skies and the oceans, the stars and the marvel of light, plants and trees, animals and as a crown on the creation, man. A striking feature of this account is that at the beginning the world is in a state of complete disorder called in Hebrew "tohu vavohu" (which in French gave "tohu bohu") and evolves progressively towards a state of increasing order — we would say from a state of high entropy towards a state of low entropy. In a biblical perspective, the question today is whether mankind will or will not preserve this state of low entropy that it has inherited and that it needs for its own survival.

2.1. The billion year time scale

Datation methods — methods that allow us to evaluate the age of formation of an object — are the indispensable tool to retrace the history of the earth and of the biosphere. The earth is 4.5 billion years old, and it is believed that life appeared on it 3.8 billion years ago. Early organisms, for instance bacteria, lived in shallow waters near vents that provided heat and minerals. By comparison, the period that we call historic times is only 10,000 years old. This time scale of civilization is 400,000 times shorter than the time scale for the apparition of life. This is like one second compared to about 4 days.

Thus many things happened to living organisms before our historical times, and it is of some interest to know about them, because they have made our present civilization possible.

2.1.1. *The apparition of life*

The connection between life and entropy is a delicate one, which has kept and still keeps many researchers busy and it is beyond the scope of this little book to dwell on it in a detailed way. We shall simply accept the empirical evidence which is that in its exchanges with the external world any living organism functions pretty much as our household does: it takes in energy and ejects out entropy. In early life forms energy was supplied as heat to organisms living under water. At that time the atmosphere consisted of gases released from the interior of the earth which was then very hot. These gases included carbon dioxide, steam (water vapor) ammonia and methane, which we call greenhouse gases. There was no oxygen, and of course no ozone layer. Therefore living organisms could only exist underwater where they were shielded from the intense and dangerous Ultra Violet (UV) radiation coming from the sun. Life outside of water would not have been possible because of this radiation. Should the ozone layer be destroyed today for any reason, life on our planet would also disappear because of the many mutations UV radiation would produce.

2.1.2. *Photosynthesis*

Around three billion years ago, a major step in the evolution of life occurred. New organisms appeared that were capable of using energy from solar radiation (instead of heat from under water vents) to maintain and develop life by a process called photosynthesis. This must have been a tricky step, as there was still no protecting ozone layer: the new organisms had to be exposed to light, but not too much. We can imagine that they developed in shallow waters where solar radiation could reach them but with the

UV filtered by a sufficiently thick layer of water. Algae were indeed the earliest form of plant life. Photosynthesis allowed life to spread over much larger areas than before, since light was available everywhere in contrast with under water heat vents that existed only at some special locations.

Photosynthesis is a process by which a living organism combines carbon dioxide from the atmosphere with water molecules, with the help of the light particles that Newton had imagined, which we call today photons. In this process, carbon from the carbon dioxide molecule is combined with hydrogen from water molecules to form an organic molecule (glucose) and at the same time oxygen is released into the atmosphere. Underwater living organisms such as algae use carbon dioxide dissolved in water.

2.1.2.1. *Photosynthesis and entropy reduction*

It is important for our purpose to track down changes in entropy that accompany the photosynthesis process, which occurs in two steps. First the energy of the incoming photon is used to excite an electron of a water molecule up to a higher energy state, breaking down the molecule of water into oxygen and hydrogen. Oxygen is released into the atmosphere. In a second step, hydrogen combines with carbon dioxide taken from the atmosphere to form a glucose molecule. All said, 6 molecules of water H_2O and 6 molecules of carbon dioxide CO_2 combine to form one molecule of glucose $C_6H_{12}O_6$ (later to be transformed into more complex organic molecules) and 6 molecules of oxygen O_2. Since 6 molecules of carbon dioxide have been replaced by 6 molecules of oxygen, the number of molecules in the atmosphere is not modified in the atmosphere. Hence entropy in the atmosphere is not directly affected by photosynthesis (eventually it will decrease because part of the released oxygen will combine with various elements present in rocks to form oxides, which are solids). Overall entropy is nevertheless reduced because in the liquid phase 6 molecules of H_2O have disappeared and only one molecule of glucose has been formed. In the photosynthesis process only a small fraction of the incoming solar radiation is stored in the resulting organic molecules. But this stored energy is of high quality, plants can be burned to provide heat at high temperature, or to provide food to other organisms. Compared to uncombined CO_2 and water molecules, glucose represents a state of lower entropy.

Photosynthesis reduces the overall entropy in the biosphere, consistent with the general rule that the entropy of a system can be reduced by an energy input, here that of the photons which have disappeared in the process. Of course plants die, like all other living organisms, and they eventually decompose into their constituents, releasing CO_2 back to the atmosphere. We are back to square one. In a steady state, what solar radiation does is to provide a way to maintain life, like the energy input into our household allows us to keep it in good order. In the biosphere there is a constant entropy turn over, it is reduced by photosynthesis and it increases when living organisms die. Yet, part of their remnants will undergo the long and complicated storage process that has led over hundreds of millions of year to the formation of reservoirs of fossil fuels. Storage affected a tiny little bit of the entropy turn over, but because it has extended over such long periods of time reserves are substantial. When we burn them we release entropy into the biosphere. It is this release of entropy, called today damage to the environment, dangers inherent to global warming and so on, which is the primary cause of concern.

2.1.2.2. *Photosynthesis and the green color of plants*

Incidentally, the green color of the material — chlorophyll — that absorbs the photons is a vivid testimony of the evolution of plants. As seen when it is decomposed by a prism, the solar spectrum is made of colors ranging from red to violet, with yellow and green in between. In early stages plants were confined to an underwater environment because they had to be protected from UV radiation. In this environment they had to make do with the long wave length (red to yellow) part of the solar spectrum, the only part that can penetrate under water. Once plants were able to grow inland (see below) they found it profitable to use the short wave length part of the spectrum, blue. In between yellow and blue there is green which plants do not absorb but reflect. This is how the earth turned green.

At first oxygen molecules produced by photosynthesis did not accumulate in the atmosphere but rather reacted with rocks such as silicates and formed oxides. But eventually enough oxygen was produced by photosynthesis for it to constitute a substantial fraction of the atmosphere.

2.1.3. *The ozone layer and the spread of life*

In the third billion year after the formation of earth, a third big step occurred, which was the formation of the ozone layer. With this protection, living organisms were finally able to propagate progressively outside of water. As they propagated inland, starting from the shores, more and more carbon dioxide in the atmosphere was replaced by oxygen, the ozone layer built up and the protection it offered became more and more efficient. One can consider that the formation of the full ozone layer was the final step that allowed the formation of the biosphere as we know it to day. Finally, about half a billion years ago plant life had spread out and much of the inland part of the earth's surface turned green.

2.2. The biosphere on the 100 million year time scale

Our knowledge of the history of the earth in the very distant past is sketchy because traces of what happened are scarce. The full establishment of the biosphere, half a billion years ago, marks the beginning of a period about which we have much more detailed information than for more ancient times, precisely because life became widespread. We are now on the 100 million time scale and can follow in much more detail the sequence of events, based on physical, geological and paleontological evidence. Remnants of living organisms have remained trapped in geological layers where they can be found today. This provides the link between geology and paleontology.

Geology follows the movements of continents and the formation of mountains and oceans by its own methods, for instance by rock dating. This method uses measures of proportions of isotopes because these change with time in a known way, due to radioactivity.

Paleontology studies the presence and nature of the remnants of living organisms in layers of sediments. In parallel, the composition of the atmosphere at different times can be followed by measuring the presence of different molecules of gas trapped in deep ice, dates being again provided by isotope dating (for instance by the ratio of two isotopes of oxygen, O-16 and O-18).

By combining geological, paleontological and atmospheric data, starting from 500 million years ago it has been possible to establish in some detail the different geological periods that have ensued. They are characterized by their respective prevailing temperature, humidity, carbon dioxide content in the atmosphere, and of course life forms. The period that we have briefly summarized above, where life was entirely under water because of the absence of an ozone layer, is called pre-cambrian. Starting from the Cambrian, 15 geological periods have been defined. But before we mention some of them, it is important to emphasize two main trends of this evolution.

2.2.1. *Carbon dioxide atmospheric content and temperature: the greenhouse effect*

Five hundred million years ago the atmosphere already contained a substantial amount of oxygen, but its carbon dioxide content was still high, roughly 20 times higher than what it is to day. The average global temperature had cooled down till about 10 degrees Celsius higher than it is now. The combination of a high carbon dioxide atmospheric content and a high temperature also occurred simultaneously in more recent times. The immediate explanation of this link is the so-called greenhouse effect. Although it has received quite a bit of publicity recently, it may still be useful to briefly recall here what it is.

2.2.1.1. *The infrared radiation*

When, after passing through the atmosphere, solar radiation hits the surface of the earth, part of it is immediately reflected back into it, part of it is absorbed in the photosynthesis process, and part of it is transformed into heat. The surface of earth, on the other hand, continuously emits radiation out to the cold skies. Contrary to the radiation of the sun, which we see, we do not see the radiation emitted by the earth, but it is there. That a relatively cold body can emit radiation can be felt by a very simple experiment. Hold your two hands parallel to each other in front of you and close your eyes. Then slowly move one of them up and the other one down, sliding passed each other. When you feel some heat, open your eyes: your hands are facing each other. This is a direct proof that your hands (and for that matter your entire body) radiate heat, even if you do not "see" this radiation (which is why it is called infrared). When your hands face each other they get a little bit warmer, because they feel the radiation emitted by the other one.

2.2.1.2. *Greenhouse gases*

Earth does the same thing as your body: it radiates heat. When that radiation on its way out hits molecules that can reflect it back again, the earth loses a little bit less heat and thus in relative terms warms up slightly. Gas molecules that can reflect back infrared radiation are called greenhouse gases (in the professional literature often called GHG). Carbon dioxide and methane molecules are two of the most important ones. The higher their concentration in the atmosphere, the more radiation will be reflected back, and the warmer the earth will get. In a green house, glass windows or plastic sheets play the role of the greenhouse gases. They let most of the solar radiation go through, but reflect a large fraction of the infrared heat radiated by the earth and plants inside the green house. The effect on the earth (or the green house) temperature is substantial. It is estimated that if there were no greenhouse gases the global average surface temperature would be lower by about 10 degrees Celsius than what it is now, which would bring it down near zero degrees Celsius. Most of the planet would be covered by ice. Not a nice prospect.

2.2.2. *Climate evolution and carbon storage*

Now we have in hand the main trend of the scenario that took place in the last several hundred million years. Growing and multiplying living organisms sucked out much of the carbon dioxide molecules that were present in the early atmosphere. They were transformed into organic molecules, stored part in the form of coal, oil, tar sands, oil shales and natural gas, and even in larger proportion in the form of carbonates through a process called weathering. These various forms of carbon dioxide storage are discussed in the next section. As the atmosphere became depleted of greenhouse gas molecules, the greenhouse effect became weaker and the surface of the earth cooled down. The general trend of the history of the biosphere is right there: as time went by, life spread, the carbon dioxide content in the atmosphere went down and so did the temperature.

We can even try to use geological records to estimate by how much the temperature would go back up if, say, the carbon dioxide content would double. If we assume for simplicity that the variation of temperature is proportional to that of the carbon dioxide concentration, and accept the geological estimates that the temperature went down by 10 degrees Celsius as the carbon dioxide concentration went down by a factor of twenty, then a doubling of that concentration would result in a temperature elevation by half a degree. Current estimates based on sophisticated computer simulations would rather give a temperature elevation of several degrees, but after all our very rough estimate gives almost the right order of magnitude.

This is almost too simple to be true. There are many factors that we have not taken into account. Ice ages are one of them: as time went by, temperature on the surface of the earth did not go down smoothly but went through many ups and downs, as did the concentration of carbon dioxide. Likewise, the progress in the spread of life on earth was not continuous, several mass extinctions occurred on the way. The radiation received from the sun was not constant either, as it varied in time due to astronomical reasons such as cyclical variations in the distance from earth to sun as well as in the inclination of the axis of rotation of the earth on its orbit around the sun, and variations in the plane of the earth's orbit.

However there remains the central indisputable fact that large amounts of carbon dioxide were taken out of the atmosphere and stored away in solid or liquid form, or as high pressure gas deposits. The overall entropy went down, because the entropy per molecule is smaller when molecules are in liquid or solid form rather than in gas form. The overall history of the biosphere, from the standpoint that interests us here, is that of a slow decrease of entropy thanks to the energy input of solar radiation via the photosynthesis process which made possible the spread of life and the storage of fossil fuels. Without the spread of life, the carbon dioxide content in the atmosphere would have remained high, the oxygen content low and the earth warmer. It would have been a rather dull story.

2.3. Carbon storage: carbonates and fossil fuels

Carbon dioxide removed from the biosphere underwent transformations resulting in carbon storage in two different ways: in carbonates and in fossil fuels. Carbonates were formed by the interaction of carbon dioxide dissolved in water with rocks, while fossil fuels are remnants of living organisms such as plants and various marine organisms that have not decomposed back into their original constituents but rather underwent a series of transformations, ending up as coal, oil and gas deposits.

2.3.1. *Carbon storage in carbonates on the billion year time scale*

Rocks are broken down by a number of processes, physical (like the freezing of water held in fractures which breaks down the rock by its volume expansion: we all know that ice floats on water which demonstrates that it contain less molecules than water in a given volume), chemical and biological (lichens and mosses form on rocks, bring in moisture; plant roots can penetrate concrete and break it down). The breaking down of rocks is globally called weathering. Now the CO_2 molecules contained in the atmosphere come into play. These molecules dissolve in water (H_2O) and are present in rain droplets. The combination of water and carbon dioxide produces carbonic acid anions (meaning they have a negative charge), which contain one atom of hydrogen, one atom of carbon and three atoms of oxygen and have a negative charge of one electron, HCO_3^-. In the weathering process Calcium cations (meaning that they have a positive charge due to electrons missing, in the case of Calcium two electrons, Ca^{2+}) are set free in water; carbonic acid anions and Calcium cations are dragged by rivers to the seas where they can combine on the sea floor, or at the surface of bacteria, to form neutral calcium carbonate molecules containing one atom of Calcium, one atom of Carbon and three atoms of oxygen, $CaCO_3$. These carbonate molecules can accumulate to the point where they will sink a bacteria down to the sea floor. In more advanced marine organisms the carbonate formation is useful in forming a protective shell that will also end up on the sea floor when the organism dies. In any case the carbonate accumulation will produce sedimentary layers.

Another way by which carbonate sedimentary layers can form involves the photosynthesis process by the bacteria or other marine organisms, in which part of the captured Carbon from CO_2 molecules dissolved in the sea combines with Calcium cations to form carbonate molecules at the surface of the bacteria, again sinking it to the sea floor. It is believed that the major fraction of carbonates formed on geological time scales and deposited as sedimentary layers has been formed with the "help" of living marine organisms.

These Carbon capture processes took place as soon as life appeared in the form of the simplest living organisms growing in heat vents, before they developed into organisms capable of photosynthesis. Of course the advent of photosynthesis must have accelerated the process because they allowed bacteria to proliferate everywhere. In any case, for a period of several billion years before the earth turned green, carbon capture and storage as carbonates must have already considerably reduced the amount of carbon dioxide in the atmosphere. Note that this process necessarily requires the presence of carbon dioxide in the atmosphere, the presence of water, and possibly also the presence of elementary living organisms. This is why one way to look for the presence of life on other planets such as Mars consists in looking for the presence of carbonates. NASA has placed in orbit around Mars satellites whose purpose was precisely to look for the presence of carbonates.

The discovery of carbonates would prove that water must have been present at some time at the surface of Mars, and that the existence of life on this planet is likely. Carbonates have indeed been identified on Mars, they provide part of the evidence for the presence of water.

2.3.2. *Carbon storage as fossil fuels on the 100 million year time scale*

Fossil fuels have preserved the chemical energy resulting from the photosynthesis process over hundreds of millions of years. Although only a very small fraction of the total living organisms produced over the years has been preserved in this way, fossil fuels constituted a formidable energy reservoir. Until recently, mankind has used very little of this reservoir.

Fossil fuels may represent only 20% of stored carbon (the rest being in the form of carbonates), but they are the only form of carbon that we can use. Fossil carbon storage started much later than carbonate formation as it required the existence of extended inland plant life, which started to spread only about 500 million years ago.

Here is a brief account of what went on, based on various datation methods.

2.3.3. *Formation of coal deposits: the carboniferous age*

Four hundred million years ago plants had developed to the point where they were forests of trees. The distinction between land and sea was not as clear cut and as stable as it is to day. Variations in the level of the oceans and land movements resulted in portions of the land being alternatively over and under water. Swamps were common. Trees would end up covered by waters where they underwent transformation resulting in the formation of peat deposits, the lowest usable form of carbon storage. At that time, continents were located quite differently from where they are today.

Places where large coal deposits have been found such as Pennsylvania were then in the tropics, much further south than where they are found now. Abundant rain and higher temperatures were favorable for forest growth, probably like in our present rain forests. The rate of carbon capture from atmospheric CO_2, more abundant than it is now, was high due to a higher temperature, high humidity and rainfall. Abundant rains play a multiple role in coal formation. They bring water inland which is necessary for photosynthesis to take place; then they participate in the weathering process and form the rivers that transport sediments and peat beds down to sea levels. Further sedimentary layers result in burial of the peat beds under thicker and thicker deposits. Peat beds are then under pressure from these upper layers, they progressively lose their water content and volatile elements. As time evolves they are progressively transformed into lignite, then bituminous and finally anthracite, the highest grade of coal. Later tectonic movements followed by erosion of upper layers may bring coal seams back up near to, or sometimes at, the surface where they can be exploited.

This is the Carboniferous Period, extending from 360 million to 280 million years ago — all said a period of about 100 million years when most of the coal deposits where formed. From the beginning to the end of this time frame CO_2 atmospheric content in the atmosphere went down by a factor of 5, to a level similar to what it is today, and the average temperature decreased by 10 degrees Celsius. This lower temperature does not immediately stop the formation of coal beds, because a large belt of high land now exists in the tropics region where temperatures remain high enough and rains are sufficient for the process of forests growth, peat bed formation and sediment transport down to sea level to continue. Eventually, however, an ice age puts an end to this process.

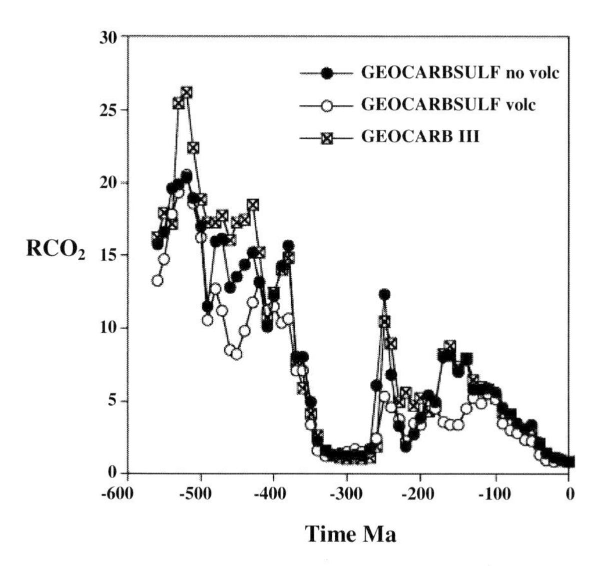

Fig. 2.1. Variation in time of CO_2 atmospheric content. RCO_2 is the ratio of that content divided by today's content. Time is obtained from Carbon isotope ratios. The graph compares raw and smoothed data. At the beginning of the carboniferous age, 360 million years ago, the CO_2 content was about 5 times higher than in the middle of that age (and today). In the interval, the temperature (not shown) went down by about 10 degrees Celsius, and is comparable to what it is today (after R.A. Berner and Z. Kothavala, American J. of Science **301**, 182 (2001)).

2.3.4. *Oil and gas deposits*

The formation of oil and gas deposits is not as well understood as that of coal. Most geologists believe that they were formed by a process similar to that of coal deposits, through the accumulation of organic matter, but from smaller organisms (not from trees), maybe primarily from marine life forms. This would have occurred on the 100 million time scale, again through compression of organic deposits under further sedimentary layers. Here however the end product is not Carbon but hydrocarbons. A competing, recent theory holds that in fact molecules composed of carbon and hydrogen were formed from deeper layers of the earth and not from atmospheric carbon dioxide. It can be called the bottom up theory, as compared to the more generally accepted top down theory.

2.4. Ice ages

When we discuss the rate at which we burn fossil fuels today it is important to remember the time frame of about 100 million years of the carboniferous age during which they where formed. In this relatively recent and better known period the carbon dioxide content went down, a shown above, by a factor of 5 (see Fig. 2.1) and the temperature by 10 degrees Celsius. If we calculate on this basis what would be the effect of an increase of the current CO_2 concentration by a factor of 2 we get a temperature increase of a couple of degrees, closer to current estimates than our first evaluation. But again this is much too simple an approach.

CO_2 removal from the atmosphere during the carboniferous age and the resulting diminished greenhouse effect were not the only driver for a decreasing temperature. Another important factor was the way continents were located. Around 300 million years ago, two major continental blocks that existed in the southern and in the northern hemisphere were joined in one single continental mass, extending fully between the two poles. It is believed that when a continental mass extends from the southern to the northern pole, as was then the case and is also almost the case today, circulation of equatorial warm waters is hindered and they transport less heat to polar regions, which get colder. Ice sheets extend further from the poles. Because they are white they reflect most of the incoming solar radiation directly back into space, and as their extension increases more solar radiation is reflected, less is transformed into heat. Earth cools down. The result is an ice age, defined as a period of time where extended ice sheets exist at the poles. According to this definition, we are in an ice age today, in spite of global warming.

Eventually, as a larger fraction of the land is covered by ice, plant growth is reduced, and less CO_2 is taken away from the atmosphere. If there is sufficient replenishment by volcanic activity its concentration can rise again, and the increased greenhouse effect will bring the temperature back up. Ice sheets will start to melt and a new cycle of plant growth can start. This is apparently what happened 250 million years ago, when CO_2 concentration and temperature went simultaneously back up. Coal deposits started again some 200 million years ago, but at a slower rate.

2.5. The last 10 million years

For the last 10 million years the CO_2 content in the atmosphere has been on the average as low as it was at the end of the carboniferous age, and for the last 3 million years the temperature, on the average, has also been similar to what it was at that time, namely rather cold. For all of that period large and thick ice sheets have been prominent on the Antarctica continent at the southern pole, and they have blocked the seas at the northern pole all year long. In spite of recent warming, the earth is still at the moment in an ice age.

But detailed records available for the last one million years show that there have been substantial variations in the climate during this ice age. There have been strong and quasi-periodical variations of temperature, CO_2 concentration and thickness and extension of the ice sheets. While the average global base temperature has been about 6 degrees lower than what it is today, interglacial periods have occurred every 100,000 years, each lasting on the order of 10,000 years during which the temperature was about the same as today. The earth is right now in such an interglacial period.

During this one million year period, variations of CO_2 concentration and temperature are synchronous, CO_2 concentration being low when the temperature is low. However, unlike what happened during the carboniferous age, it is unlikely that there is a causal relationship between the two, namely it is unlikely that the periods of lower temperature are due to a reduced greenhouse effect. One reason is that the variation in CO_2 concentration is too small to explain the amplitude of temperature changes, about 6 to 10 degrees. The second is the clear periodicity of interglacial periods, which points out to an astronomical effect. Periodic variations of the eccentricity of the elliptical orbit of the earth, of the inclination of its axis of revolution and of its precession as well as periodic variations of the angle of the earth's orbit with respect to the plane of the solar system as a whole, as well as combinations of the above, which all produce periodical variations of the solar radiation received by the earth, have been proposed as being possibly at the origin of the periodical interglacial periods. While details still remain to be worked out, it is believed that the succession of alternate glacial and interglacial periods is due to changes in the incoming solar radiation and its complex interaction with the bio-sphere. This is discussed in more detail in Chapter 5.

The current interglacial period started about 10,000 years ago. It is during that period of time that civilizations have flourished on earth, based on sedentary life made possible by the development of agriculture, itself made possible by the stable climate. At the start of this period, rising sea levels must have changed the landscape considerably in coastal areas, possibly wiping out settlements as recounted in the story of Noah and in many other mythological stories of a number of civilizations. Possible changes in climatic conditions have since been a constant worry, as they would endanger our current civilization in which we get almost all our food from agriculture, rather than from hunting or fishing as did our farther ancestors.

The strong decrease of the CO_2 content in the atmosphere, on the 100 million year time scale, has definitely played a major role in the general earth's cool down. Together with the motion of the continents, it has led to an overall tendency to ice ages, but what has made our current civilization possible is an interglacial period. Various views on how long it will last are reviewed in Chapter 5. We are not sure when it will end, if it is in the next 1000 years or 10,000 years or more. But it will, and this is one of the major concerns for the continuation of civilization itself in the long term. If glaciers were advancing again in Europe and North America, not much would be left of our towns and overall infrastructure in these regions.

It is against this background that we will consider the possible impact of global warming, possibly due today to the increasing level of atmospheric CO_2, but before we get to that point more quantitative discussions of energy and entropy are necessary.

Chapter 3

How Much Energy do We Need?

One of the basic laws of physics is that energy is conserved, but unfortunately, we do not have an intuitive understanding of what energy is. Concepts such as heat and force are obvious to us, because we directly feel heat and force. We also measure speed easily. Another quantity that is conserved is momentum, the product of speed and mass, and we can experiment with this law when we play billiard; but the concept of energy and of its conservation is really only useful when energy is being transformed from one form to another. Newton himself, giant amongst the giants of Science, did not have a clear notion of what energy is. He thought that the energy of a moving body, which we call kinetic energy, is proportional to its speed, while it is proportional to its square, as rightly claimed by Leibnitz.

While energy is undoubtedly conserved when transformed from one form to another, it is our common experience that this is not the whole story. If the total energy was all that mattered, we could not have an energy crisis, so we know that some essential concept is missing from the conservation statement. For instance, we know that we can transform mechanical energy into heat with excellent efficiency (in fact 100%) as when we use breaks to slow down our car, but we also know that the inverse is not true. Heat cannot be transformed back into mechanical energy with 100% efficiency. Heat collected in solar water heaters can be used to produce vapor from a fluid that will activate a turbine, but the conversion efficiency will be very low. It was Carnot who first showed that the efficiency with which heat can be transformed into mechanical energy, like in a steam engine that operates between a high temperature T_H (that of the boiler) and a low temperature T_c (that of the condenser), can never be higher than a limiting value. That value is equal to the ratio $[(T_H - T_C)/ T_H]$. It is reached for a particular cycle — the Carnot cycle — during which the entropy release is zero A more quantitative discussion of entropy will be found in the next chapter.

3.1. Different forms of energy and power

In order to discuss usefully our energy requirements, we must first consider the different forms of energy that we use and conversion possibilities from one form to another, heat and mechanical energies being only two examples amongst several more forms.

Table 3.1. Different forms of energy and possible conversions. Devices that can convert different energy forms into one another. See text for details and possible conversion efficiencies.

From/ Into	mechanical	heat	Electrical	Chemical	Radiation	food
Mechanical		Friction	Generator			
heat	Heat engine		Thermo-electricity			
electrical	motor	Resistor		batteries	Antenna	
chemical			Fuel-cells batteries			
radiation		Solar thermal	Photo-voltaic	Photo-synthesis		plant
Food, biomass		100%		biofuel		

Table 3.1 lists the main forms of energy that we use, and gives some information on energy conversion possibilities. The main forms of energy are: mechanical; heat; electrical; chemical; electro-magnetic radiation; nuclear; food.

Power is the energy spent divided by the time over which it has been spent. In order to allow comparisons between different forms of energy, we must use the same system of units for all of them. For practical reasons, power is often used as the primary quantity, and energy is expressed as the power spent multiplied by the time over which it has been spent. Power is usually expressed in units of kilowatt, and energy in units of kilowatt-hours. We are familiar with kilowatt-hours, or kWh, because this unit of energy is used to calculate our electricity bill. This is the unit of energy that we shall use for our comparisons.

Mechanical energy is the work that we produce when we move a body at constant velocity over a length L by applying a force F to it, if the force is applied along the direction of the displacement. A good example is that of a horse carriage. The horse applies a certain force to the carriage, and if the road is straight the product of that force by the distance over which the horse has moved the carriage is equal to the work produced by the horse. For historical reasons, mechanical power is sometimes expressed in units of "horse-power". One horse-power is equal to 0.736 kW. That power is of the same order as that of a one-room small air conditioning unit. The engine under the hood of a typical car can develop 100 horse-power, or 74 kW. If we drive it for one hour at full power, or for two hours at half power which would be typical for daily commuting to work, we will have spent 74 kWh. Incidentally, these 100 horses under the hood of our motorcar are a good illustration of our modern way of life. Not long ago, only kings could afford to own 100 horses and their ancillary equipment (stables, care takers of different kinds and so on). To day, in the developed world, everybody has them and in the developing world everybody wants to have them.

Heat energy is still often expressed in units of calories. One calorie is the amount of heat necessary to heat one gram (or one cubic centimeter) of water by one degree. It is equivalent to 4.18 Watt-second. To heat up 150 liters of water by 50 degrees, which is roughly what a family of 4 may need for its daily domestic hot water use, we need 7,500,000 calories, or 9 kWh.

Electrical energy is familiar to us as the energy spent in appliances such as air conditioning units, refrigerators, vacuum-cleaners, mixers, light bulbs, furnaces, washing machines and dish-washers. In these various appliances, electrical energy is easily and efficiently converted into mechanical energy, radiation and heat. Just as heat flows from hot to cold regions, electricity flows from high to low potential. Electrical power spent is equal to the flow of electricity measured in Amperes (A) multiplied by the difference in potential, measured in Volts (V). When a current of one A flows down a potential difference of one Volt, a power of one Watt is being dissipated (the flow of electrical current is that of electrons, each of them carrying a very small electric charge denoted by *e*). The power dissipated in a light bulb or in a personal computer is of the order of 100 W, while the power dissipated in heavier appliances such as refrigerators, air conditioning units and so on is of the order of 1 kW. It has been estimated that the average electrical energy spent on appliances in a household is of the order of 10 kWh per day (in Europe).

Chemical energy is a term which we use here in a broad sense. It covers not only the energy stored in fossil fuels (coal, oil, oil sands, natural gas), and biomass, but also hydrogen and energy stored in batteries. The commonest way to recover chemical energy is by combustion, for instance through the reaction $C + O_2 \rightarrow CO_2$. Energy is recovered in the form of heat because this reaction is exothermic. The energy recovered is that of the chemical bonds between the carbon and oxygen atoms, broken in the combustion process. It is of the order of one electron-volt (or one eV) which is the energy one must spend to elevate the potential of one electron by one Volt. To get an order of magnitude, the combustion of one liter of liquid fuel releases about 10 kWh. But hydrogen can also be converted into electricity in a fuel cell where it is recombined with oxygen to form water, and chemical energy stored in a battery is directly recovered as electrical energy.

Radiation energy is the form of energy that we receive from the sun. For each wave length of the solar spectrum, energy is transmitted in a quantized way in the form of photons (the "grains of light" of Newton), each of them having the energy:

$$\varepsilon = hc/\lambda$$

where h is a universal constant (Planck's constant), c is the speed of light, and λ is the wave length of the radiation. h is in units of energy times time. Expressed in units of kWh, or even Wh, the energy of a photon is extremely small. It is more convenient to express it in units of electron-volt, denoted above as eV. When we calculate the energy of one photon in the visible part of the solar spectrum, which has a wave length of the order of 1 micron, we find that it is of the order of 1 eV. This is the order of magnitude of the energy needed to break up a chemical bond, which is the reason why photons can trigger chemical reactions. This is what occurs in photo-synthesis, as we have discussed in the previous chapter.

Nuclear Energy is produced when a heavy unstable nucleus is split into two parts whose total mass is slightly smaller than that of the original nucleus. The value of the energy produced is given by the famous Einstein equation:

$$E = mc^2$$

where m is the mass lost in the reaction. In a typical nuclear fission event such as occurs when the Uranium 235 isotope (meaning the isotope of Uranium that has an atomic mass of 235 units, the hydrogen nucleus having a mass of 1) is split (for instance when hit by a neutron), the energy released is of the order of tens of M*e*V (1 MeV is one million eV). The energy is released in the form of kinetic energy of the reaction products which are emitted with a high velocity. In a nuclear reactor this kinetic energy is then transformed into heat, which will be further converted into mechanical energy and electricity (which is why we have not listed nuclear energy in Table 3.1). It will be appreciated that the energy obtained from a fission event is about 10 million times larger than in a chemical reaction.

Food energy is really a form of chemical energy, but we give it a special mention because after all it is for us the most essential form of energy. We can eat wheat and corn, vegetables, fruits, meat — but we cannot feed ourselves on coal, gasoline, or electricity. Food energy is produced from solar radiation by photosynthesis. As we have seen in Chapter 1, a photon breaks up a molecule of water, oxygen is released into the atmosphere and hydrogen is combined with CO_2 molecules from the atmosphere to form carbohydrate molecules. The way our body burns food is by breaking up these molecules and eventually releasing CO_2 and water molecules. Back to square one.

In many publications, food energy does not appear amongst the list of our energy needs. From which one may be tempted to conclude that it is negligible compared to other energy needs. In fact this is not quite true. Based on the rate at which CO_2 molecules are released by our body, it has been calculated that the average power dissipated by an individual is of about 100 W to 150 W depending only slightly on the amount of physical work he/she produces, giving an energy spent of about 3 kWh per day. For a family of four, this represents an energy expenditure of the same order as the electrical energy used for appliances. *Food energy can represent close to a quarter of the total energy spent in the framework of a household. This figure varies of course with the socio-economical level of the particular household considered. It will be less for well to do families because their house is bigger and uses more appliances, and larger for lower income families. For more details, see Section* ***3.4.***

3.2. Energy conversion

Table 3.1 gives the names of some devices used for energy conversions. Here we indicate possible conversion efficiencies.

Mechanical energy can be converted into heat and more importantly into electrical energy at close to 100% efficiency, as is done in high efficiency electrical generators in large scale power plants.

Heat energy can be converted into mechanical energy within the limits imposed by the Carnot law, discussed in some detail in Chapter 4. In modern power plants, this conversion efficiency exceeds 40% and can approach 50%. The conversion efficiency of an internal combustion engine is much lower, about 20%. This means that 80% of the energy content of petrol is lost in the engine. Direct conversion of heat flow into electrical power is possible but not very efficient. A hot body emits radiation as we can see when we burn wood in our chimney — or when we sun-bath.

Electrical energy is very versatile. It can be converted with high efficiency into mechanical energy by electrical motors, into heat (like in an electrical resistance), into chemical energy (like in a battery or by electrolysis to produce hydrogen), and into radiation (like in a microwave oven).

Nuclear energy, by contrast with electrical energy, can only be converted into heat, at least in the present state of technology.

Chemical energy is usually first converted into heat, except in fuel cells and batteries where it can be converted directly into electricity as mentioned above. Fuel cells perform the reverse operation from electrolysis, hydrogen being recombined with oxygen to form water while electricity is generated. This is the basis of what has been called the "hydrogen economy". Of course hydrogen must be produced in the first place, for instance by electrolysis. All said, hydrogen is used in fuel cells as an agent to effectively store electricity. The advantage over conventional storage by batteries is that the energy stored by unit weight is much higher, because hydrogen is the lightest of all elements. However hydrogen is a gas that must be stored in some condensed way other wise it would take too much space. This entails the considerable additional weight of high pressure containers, except if hydrogen is stored in liquid form as it is in space rockets.

Radiation energy is as versatile as electrical energy. In fact, we could classify both energy forms under the common heading of "electro-magnetic energy".

The energy of photons is converted into heat when they hit a body that absorbs the radiation (a black body). This conversion is used for instance in solar water heaters.

In photovoltaic cells (PV cells), photon energy is converted into electricity. Incoming photons excite electrons from a lower energy state to a higher one, leaving behind them a “hole” that has the opposite electrical charge (positive). A built-in electric field then separates the electron from the hole. They move in opposite directions, giving an electrical current. The most advanced laboratory photovoltaic cells approach a conversion efficiency of 40%. Commercial panels approach an efficiency of 20%. Calculated on the basis of a day and night global average, incoming radiation from the sun on a horizontal surface in inhabited regions amounts to 200 W per square meter. With cells having 20% conversion efficiency, the available average power would be 40 W/m^2. All commercial cells have conversion efficiencies higher than 10%.

On the geological time scale, solar radiation energy has been transformed into chemical energy in the form of fossil fuels, as we have seen in the preceding chapter. Until now these fossil fuels constitute our principal primary energy source. They are not renewable on our time scale, since their formation took on the order of 100 million years.

Biomass energy. On a renewable basis, we can of course burn plants, either directly or after converting them into liquid fuel. This is called biomass energy. Although the primary conversion efficiency in the photosynthesis process is high — about the same as in photovoltaic cells — the effective efficiency of biomass production is in fact quite low because some of the incoming radiation is reflected, and because plant growth requires respiration which requires energy. The theoretical efficiency is therefore limited to 3 to 6%, which corresponds to 6 to 12 W/m^2 on an average basis. In fact the practical efficiency as measured in crops is even lower and gives only about 1 W/m^2 average power. Energy needed for planting and harvesting further reduces this figure to about 0.5 W/m^2, or half of one percent of the incoming radiation energy. This is 60 times less than photovoltaic power using commercial cells having 16% conversion efficiency. In other words, biomass requires 60 times more land than photovoltaic cells to produce the same amount of energy.

Conversion of radiation into food is of course of primary importance for us. Food production uses solar radiation but in principle we could use radiation from a different source, provided it had the appropriate wave length. For instance, we can think of a spatial vessel sent on a very long journey into space at the edge of, or even far away from the solar system, where electricity would be produced by a nuclear reactor and transformed into radiation making "spatial agriculture" possible for the cosmonauts. Similarly, "underground agriculture" is also possible. Supply of water and carbon dioxide molecules, as well as oxygen, are not a concern since the system would operate in a closed cycle.

We give these examples not just for fun, but to make an important point. This spatial vessel or underground station where life can be sustained with the help of an energy source — here nuclear — are conceptually not different from our biosphere. It is also a closed system, except for the input of solar radiation. Maintaining life requires an energy input. A discussion of how much is needed is precisely the object of this chapter.

3.3. Energy use and entropy release

In the end, energy that we use in all its forms is generally transformed into heat, often called waste heat or low grade heat. This heat, rejected into the environment increases its entropy.

3.3.1. *Heat rejection*

Let us consider for instance the use we make of electrical energy. We have given a few examples of how it can be produced, but after a number of transformations heat will always be the end result. This is what happens in our household. In most light bulbs a large fraction of the power is directly dissipated in the form of heat. Then the radiation (light) itself will be eventually absorbed by the environment (walls and so on) and transformed into heat. The same occurs in TV and radio sets. A refrigerator produces heat, part of it is of course that ejected from its inside into the kitchen, but heat is also generated by the electrical motor and the compressor it drives. The same is true for air-conditioning units.

Energy spent on transportation all ends up as heat, whether we use chemical energy in cars, buses or airplanes, or electrical energy in trains. The heat produced in our car's engine is directly wasted into the atmosphere. The mechanical work that it produces eventually ends up as heat by friction against the atmosphere, tires deformation, and internal friction in the transmission.

But there is one interesting exception to the general rule that needs to be mentioned. This is when energy is used to produce materials that later on can themselves be used to produce energy. For instance, the manufacture of photovoltaic modules requires the production of materials that are not found in nature, such as pure silicium, glass, aluminum and so on. This production requires energy: just as is the case for agriculture, harvesting the sun first demands an energy investment.

As one last example of heat rejection let us go back to our own energy needs. Again, the 100 W or so that our body dissipates end up as heat: our body operates at 37 degrees centigrades, a temperature higher than that of the environment under normal conditions (perspiration can help us survive in slightly higher temperatures). We take in energy in the form of food (chemical energy), and eject it as heat into the environment. This occurs even if we do not perform any physical work. Why is that? Clearly, we are different in this respect from all the equipment we use. When appliances are switched off, no energy is spent.

Yet, if we consider ourselves as a subsystem of a larger one (let us say of the household, which is itself a subsystem of the society at large), we do see that at all levels we operate in similar ways. We understood in Chapter 1 that an energy input into the household is necessary just to keep it in good order. We do not turn on the vacuum-cleaner arbitrarily: we need to use it in order to keep the house clean. The same applies to us. Even if we do not produce any physical work, an energy input — here in the form of food — is necessary just to keep ourselves "in good order". We, as our household, are systems that need to be kept in a properly functioning steady state. The consequence of this up-keep is that we — ourselves, our household and society at large — eject low grade heat into the external world, and in addition we also pollute it.

3.3.2. *Entropy release*

In the language of physicists, ejection of low grade heat and pollutants into the environment amount to release of entropy. All, or nearly all (see the case of photovoltaic cells), the energy that mankind consumes ends up as an increase of entropy in the biosphere. The only counterweight to that increase is photosynthesis, which decreases the entropy of the biosphere if part of plant production is eventually stored, as it has been in the past, in the form of fossil fuels.

We are now almost in a position to estimate what the overall entropy balance is, which we will do a little bit later. It suffices to say at this point that in modern times there is a net increase in entropy of the biosphere. From that standpoint, it does not really matter whether we burn fossil fuels, nuclear fuel or biomass.

What we need now are a few quantitative definitions to justify the above statements.

3.4. Energy needs and costs

We are now equipped to evaluate and compare our different energy needs and see how they could be covered. Some of them vary enormously from one country to another, others don't.

3.4.1. *Food energy*

Food energy requirements are basically the same for all humans. They do not vary substantially with race and conditions of living. Even the amount of physical work produced will not affect them by more than about 30%, except in extreme cases. Another universal constraint is that there is basically no choice as to the form of energy used to produce food: there exists at present no alternative to solar radiation and photosynthesis. As already said, we cannot feed ourselves on coal, oil or electricity.

The power that we dissipate, of the order of 100 W, and the corresponding energy needed per day of about 3 kWh, are relatively modest in regard to other energy needs in developed countries. But they are not at all negligible in third world countries. For instance, the total average power used per person in Canada is 14 kW, two orders of magnitude more than the power that our body dissipates. By contrast, in Nigeria the average power consumption per person, excluding food, is only 43 W. In that country more power is consumed for food than for all the other "needs" lumped together. The worldwide average power consumption per person is about 2.3 kW. In view of the enormous difference between the power consumed per capita in developed and third world countries, this average does not have much meaning. Hundreds of millions of human beings consume much more, and billions much less.

From an economic standpoint, the price of a kWh of food is much more expensive than a kWh of say electricity, see Table 3.3 for some examples. There are a number of reasons for that, many of them related to dietary habits which can vary considerably according to the standard of living. For instance, a kWh of meat is evidently more expensive than a kWh of rice. But the high cost of food energy has also a fundamental reason. Food is basically produced by photosynthesis, which has a low yield compared to photovoltaic conversion, as mentioned above. Now remember that photovoltaic electricity is itself expensive compared say to that produced by burning conventional fuels, roughly by a factor of 4 today. It is therefore no surprise that even a kWh of basic food (cereals, rice, bread) is more expensive than a kWh of electricity.

This has immediate economic consequences when it comes to the budget we live on. While the 100 W that we dissipate are negligible compared to the total power consumed in developed economies, their cost is not. In middle class families it can represent from a quarter to a half of their total budget. But in third world countries, it is close to 100 percent.

Therefore, we must be very careful not to forget food energy needs on a global scale. This is in fact a very sensitive issue. For instance, replacing petrol by biofuels may appear as an attractive proposition in developed countries, possibly for political reasons, but if this replacement were to take place on a large scale, the resulting reduction in food production and price increase would I believe mean hunger for hundreds of millions. A doubling of the price of basic foods (cereals, rice) has already occurred in recent years. Massive production of biofuels on a large scale in the Americas and in some countries in Europe is one of the factors behind this increase. It will make people unhappy everywhere but will not much affect those who live in developed countries. For most people in the third world, it might mean starvation.

3.4.2. *Food versus other energy needs*

We accept as a fact that our basic food energy needs — those 100 W that we dissipate no matter what — are ireducible. And in fact they are usually not even listed in most publications that deal with the "energy problem", since no saving is possible there. On the other hand, we consider that we can save on our other energy needs. Maybe this difference in attitude is worth a little digression. This will take us right back to entropy.

As all living organisms we need energy just to keep ourselves alive, irrespective of the work we do. Living organisms are not stable if left to themselves, contrary to inanimate forms such as crystals. In crystals, atoms are fixed in space for eternity unless some external agent attacks them, but in living organisms atoms and molecules are in a constant flux. The organism may look the same all the time but in fact if we were to look at it on the microscopic scale we would find that molecules are all the time exchanged. Cells die and are replaced by new ones. Entropy is all the time released, just as in the example of the household that we discussed in Chapter 1 and, just as in the household we need an energy input to keep this increasing entropy in check. The entropy generated in living organisms is dumped in the external world, again just as in the case of the household. Without an energy input, living organisms will die just as our house hold will disintegrate. The 100 W that we dissipate is a measure of how much power input we need to keep ourselves in a good state of repair. Perhaps the most striking example is the brain. It consumes around a fifth of our total power dissipation, while evidently not doing any work in the usual sense: it does not move at all. Its dissipation level of about 20 W is the result of a long evolution. When compared to the dissipation of personal computers, which is of the order of a few 100 W, its performance is truly impressive: it does so much more.

3.4.3. *A family's energy needs*

After this digression it is time to go back to our different energy needs. Instead of looking at global numbers at the society level, we shall take the same approach as in Chapter 1 — looking at a household.

Let us take the example of a family of four living under one roof. Besides food energy, it consumes energy in the form of electricity for appliances, of electricity, or fuel, or natural gas for heating, and of petrol for car transportation. In order to compare the different components of its energy budget, we use the same unit — kWh per day — for these different kinds of needs. The numbers in the following table are indicative for a family living in a moderate western climate, let us say Switzerland to illustrate the point.

Numbers in Table 3.2 will of course vary from family to family according to their standard of living, how well their home is thermally insulated, how many cars they own and how much they use them and so on. The reader can easily go through the little exercise of drawing a more accurate table for their own case. What is important here is to note the order of magnitude. From the total energy consumption of the house hold per day — 110 kWh — we can calculate the average power spent per person. It is of about 1 kW. It is important to stress that this number represents the power spent at the household level, not the total power spent.

Keeping these caveat in mind, we can draw from this table a number of interesting conclusions, valid for developed countries. Our food energy needs represent about 10% of the energy needs of the household. They are on the same level as the energy we need for our appliances. The major items — and this is not a surprise — are for heating and transportation (even though the modest number of kilometers traveled in the family car given in the table evidently does not include the use of other, public transportation).

3.4.4. *A family's energy costs*

We compare briefly below food energy and other energy costs.

Table 3.2. Household daily energy needs for a family of four in kWh.

Food energy Based on an average power consumption of 150 W, taking into account a modicum of physical activity	**10**
Electrical appliances Includes refrigerator, washing machines, television, personal computer, light etc…	**10**
Hot water For sanitary use	**10**
Home heating For a well insulated single family house of 160 sq. meters where 1000 liters of fuel or equivalent are consumed per year	**30**
Transportation Corresponding to an annual driving of 15,000 km driven in a car consuming on the average 10 liters of petrol for 100 km. There are 10 kWh of stored chemical energy in one liter of petrol.	**45**
Total	**110**

3.4.4.1. *Food energy costs*

Table 3.3 gives the price we pay for a kWh of selected foods in a typical Euroland location. It ranges from 0.3 Euros/kWh for spaghetti to 17 Euros/kWh for beef, while the price for a kWh of electricity is around 0.08 Euros. One kWh of spaghetti is four times and 1 kWh of beef 200 times more expensive than one kWh of electricity. Even if we had no electrical appliances at all, the corresponding electricity savings per day would just be enough to feed one person on spaghetti or rice.

Table 3.3. Typical costs of a kWh of selected foods in a Euroland country, in Euros (2007).

Rice, spaghetti	0.3
Bread	1.1
Potatoes	1.0
Tomatoes	12
Chicken	8
Beef	17

3.4.4.2. *The different energy costs*

Table 3.4 shows how food, a relatively small item in the family's energy budget, becomes a dominant one when we consider its cost, compared to that of other energy expenses. Numbers given here are only indicative, they will vary for a number of reasons for instance because the price of 1 kWh of electricity will easily vary by a factor of 2 amongst Euroland countries and of course because different dietary habits and the type of local food availability will also vary. These variations will be more pronounced amongst different continents, but the overall conclusion that a kWh of food is several times more expensive than a kWh of electricity or of any of the other forms of energy we consume is quite general and holds everywhere.

One may wonder why food energy, a minority item in the energy spent at the household level, is the dominant one in terms of energy-related expenses. One reason is the low yield of the photosynthesis energy growth process. Even if we were to obtain all of our electricity needs from photovoltaic conversion, which would increase the price of one kWh of electricity by a factor of about 4, the price of 1 kWh of electricity would still be less than the price of one kWh of food energy.

The low yield of the photosynthesis process is compounded by the fact that in modern societies we get a substantial fraction of our food energy from meat rather than from plants. The high price of beef energy, see Table 3.3, reflects the very low efficiency of transforming grain energy into meat energy. If we take the price ratio between grain energy (wheat) and beef energy as an indication of the conversion efficiency of the "beef machine', we get that it is only of a few percent. This low efficiency implies that in the conversion process from grain to meat most of the energy stored in grains is released in the biosphere as entropy. As pointed out in an article that appeared on January 27, 2008, in the New York Times, greenhouse gases released from food production in meat eating societies may be of the same order as that released by cars.

Table 3.4. Daily household energy budget for a family of four in Euros.

Food energy (minimum)	**15**
Electrical appliances	**1**
Hot water	**1**
Home heating	**4**
Transport	**6**
Total	**27**

In a household, food costs represent the dominant part of the energy budget, even if the family is fed mostly on spaghetti, potatoes and bread. Therefore, and this is our main point, it should be mandatory to include food energy in an overall energy strategy. Up to now, this has not been done. For instance, there is no reference to food energy in the statistics published by the International Energy Agency. Because of the recent and rather massive development of biofuels, food expenses and fuel expenses for transport are no longer independent quantities since food and biofuels compete for land. Attempts at reducing the transport budget line by promoting biofuels through tax incentives make little sense because such policies will result in an increase of the food energy budget line, which will be much larger than savings on transport and will be most damaging to low income families. Mass production of biofuels would be a socially irresponsible policy.

3.4.5. *Energy needs at the society level and the entropy problem*

As we have seen above, the total power spent in a household corresponds to an average power consumption of about 1 kW per person. In a typical western society, the total power consumption per person is several times larger, it may be 5 kW in Western Europe and more like 10 kW in Northern America.

Thus, in western-style countries, the household energy needs per capita are only a fraction of the total average energy spent. But total power consumption per capita varies a great deal from continent to continent and from country to country. Qatar seems to have the highest per capita consumption — 29 kW — and Senegal has one of the lowest, 0.3 kW. A typical figure for large emerging economies such as China and India is of the order of 1 kW (1.6 kW in China, 0.7 kW in India).

Modern societies are evidently extraordinarily complex systems. We may consider them as a form of living entity. Just like our body does, modern society needs some kind of "food energy" to keep itself going.

Our body looks the same from one day to the next, but it is not actually quite the same since cells have died and have been replaced, others have been repaired. Chemicals that could potentially poison it have been eliminated. This occurs thanks to the food that has been taken in and digested. In other words, entropy has been maintained at an acceptable level thanks to the intake of food energy and entropy release into the external word.

Similarly, cities look the same from one day to the next although they also are not quite the same. They need constant repair and maintenance. The large power that modern societies consume, and over which individuals have no direct control, is just the consequence of the necessity to keep their entropy at an acceptable level. This entropy, just as is the case for living bodies, is released to what we call the "environment". The oil and coal that are burned at the society level constitute the "food energy" that our societies need to stay alive. Pollution is the result of the entropy that they must eject.

Most of the energy consumed in a developed country is indeed "burned" in the general infrastructure, places of work, industry, schools, other public buildings and so on, and not at the household level. It is not under the control of individuals. This is why it is so difficult to achieve substantial energy savings by individual decisions. We are all encouraged to save energy, but in fact in our private sphere there is not very much we can do. Of course, we can invest in a better house insulation, in more energy efficient light bulbs and in a car burning less gas, and certainly we should do so. But on the level of society as a whole, this will only translate in limited savings.

To go back to the example of our household, life in the society we live in *requires* that we go to work regularly, children *must* go to school, maybe if both spouses "work" a cleaning person *must* come and go back to his/her own living quarters. When at work, we engage in actions that our business requires, which will affect the actions of other people and so on. In spite of all this "work", after a 24 hours cycle everything will look pretty much the same as it did the day before. In a strict physical sense we have not accomplished any work, we have just engaged in our daily fight against entropy and in so doing consumed the required energy and contributed to polluting the environment.

This is why what is called an energy crisis is actually an entropy crisis. It has two aspects that cannot be dissociated. The first is that the amount of energy needed to keep the entropy increase in check may surpass the energy actually available. The second is the increasing pollution of the environment at the level of the biosphere. It is precisely because we are dealing with an entropy crisis that these two aspects are so intimately connected.

3.5. Can society survive with a lower entropy release?

Of course, it can.

A first and radical option is for people to stop driving cars, flying airplanes, using electricity, in one word go back in time. This is the approach of some extreme ecologists.

Another option is to keep living the way we do now, but drive cars and fly airplanes that use less fuel, use more efficient appliances, improve thermal insulation of private and public buildings, and so on. This is what is going on right now.

A third way is to modify the way society functions, for instance do much more work at home, discourage secondary homes, encourage proximity shopping, schools, hospitals and services in general. This implies changing our way of life. This will have to be done eventually because this is how society can best reduce the amount of entropy it produces and must get rid of.

The challenge we face is to develop post modern societies that will function in such a way that all of humanity can participate in them. We cannot avoid this challenge, because the amount of entropy produced by modern societies is so large that if all countries were to join in their way of living there would not be enough energy sources to get rid of it and we would all choke under massive air and water pollution, not to say anything (yet) on the problems posed by climate change.

This challenge is what we are facing right now. Let us look at the numbers again. If the power consumed per capita was to be the same everywhere as it is say in Northern America, energy consumption would increase from about 2 kW (the current world average) to 10 kW per capita. We would need to burn 5 times more oil and coal than we do today. These resources simply do not seem to exist. Within 5 years we would have exhausted oil reserves, within 30 years coal reserves. Pollution would become colossal, CO_2 emissions would quintuple. Even the warmest proponents of solar energy will recognize that such a level of energy production cannot be maintained. Yet, the desire for the "good life" is now irresistible as it spreads fast in China, India, Asia and will also spread soon in Africa.

The way forward is to recognize that modern societies are really in their infancy — they only started to develop 200 years ago or so. Human beings, on the other hand, are the result of evolution over millions of years and indeed, entropy wise, we are, as individuals, tremendously effective. Consider what we can do while dissipating a mere 100 W. At the household level we already consume 10 times more power than at the individual level (for food only). At the society level we consume again 10 times more power than at the household level. A society that dissipates per capita 100 times more than the individuals, just to keep going, must be considered as very primitive. The challenge is to develop post modern societies that will dissipate perhaps 10 times less.

An encouraging example is the evolution of computers. In terms of computing power the equivalent of today's personal computer would 50 years ago have filled an entire apartment and gobbled up orders of magnitude more power. The way forward is to use more science and technology, not less. It is also to better understand what societies are all about and how they can be modified to reduce their tremendous entropy production.

Chapter 4

Entropy in Thermodynamics and Our Energy Needs

As the name indicates, thermodynamics deals with heat and motion. This branch of physics developed after the invention of the steam engine which transforms the heat of the steam coming from the boiler into mechanical energy. This transformation from heat into mechanical energy occurs via the pressure that steam exercises on a piston which it gets moving inside a cylinder. After it does its work the steam is released and condenses back into water. In the process, heat has been transferred from the hot boiler into the colder environment, and work has been performed by the moving piston

This immediately poses the question of the equivalence between these two forms of energy, heat and mechanical energy, and of the efficiency of the engine in converting the first into the second. Here entropy plays a decisive role, which we will discuss in the first part of this chapter. In particular, we shall see how the increase of entropy in the course of the conversion process must be compensated by an input of energy, a notion that we introduced already, qualitatively, in Chapter 1.

4.1. Entropy in thermodynamics

4.1.1. *Heat and mechanical work as two forms of energy: the first law of thermodynamics*

While the transformation of heat into mechanical energy is a relatively recent invention with the discovery of the heat engine (and later the internal combustion engine), it has been known to mankind since the dawn of civilization that mechanical work can be transformed into heat. Today we light a match by rubbing it against a rough surface, the heat produced by friction raising sufficiently the temperature of the tip of the match to put it on fire. Life was more difficult for our ancestors, but they used the same physical principle apparently by rotating rapidly a sharpened tip of hard wood against another piece of wood. Another striking (and potentially dangerous) example is the strong rise of temperature that occurs on parts of a space shuttle as it re-enters the atmosphere. Friction of the shuttle against the atmosphere reduces its speed, and the loss of energy of motion (which we call kinetic energy) "reappears" as heat that warms up dangerously the exposed surface of the shuttle.

Heat and mechanical energy can thus be transformed into one another — they are two forms of energy. The first law of thermodynamics states that energy is conserved. When transformations occur between different forms of energy — and besides heat and mechanical work there are several more forms of energy as we have seen in the preceding chapter — energy is neither created nor lost. We can calibrate the different forms of energy against one another to make the appropriate book-keeping and verify that this law is obeyed. For instance the amount of heat necessary to raise the temperature of one gram of water by one degree Celsius — one calorie — is equivalent to 4.18 times the energy spent when a power of one watt is dissipated during one second. This amount of electrical energy is one Joule:

$$1 \text{ calorie} = 4.18 \text{ Joule}$$

For instance, we can dissipate 4.18 Joule in one gram of water by dissipating an electrical current of one Ampere during 4.18 seconds in a resistor of one Ohm, and verify that its temperature has risen exactly by one degree (if the experiment is performed carefully, which requires that no heat be transferred to the external word). But can we transform 1 calorie into 4.18 Joule with 100% efficiency? As already indicated, the answer to this question is (unfortunately) no. It turns out, as we shall show below, that only a fraction of the heat put into the steam engine can be transformed into mechanical energy. The rest is released as low grade heat into the environment. The work done (mechanical energy), plus the heat released, are equal to the heat input, so that the law of conservation of energy is not violated, but an additional concept is necessary to predict how much of the heat put into the steam engine can be transformed into mechanical energy. To introduce this new concept, we must first take a closer look at the way the steam engine actually works.

4.1.2. *Thermodynamic cycles*

So far, we have described the transformation of heat into mechanical work by using the simple example of hot and high pressure steam introduced into a cylinder where it gets a piston moving, but this is not quite the way a steam engine works. After moving one way the piston must be brought back to its initial position, the expanded steam must be released and new hot high pressure steam must be introduced again and so on. The engine works actually in cycles. After one full cycle the piston is back to its initial position, and the temperature and pressure of the steam are also back to their initial values.

The first theoretical description of thermodynamic cycles is due to Sadi Carnot who in 1824 published it in his book "Reflections on the Motive Power of Fire". He was inspired by the work of his father, Lazare Carnot, a military engineer, who was interested in the efficiency of water-powered machines. In a perfect machine, work produced by falling water would ideally be sufficient to pump it back to its initial level. Sadi Carnot adopted the point of view of Antoine Lavoisier according to whom heat is transported by particles that he called "caloric". Sadi Carnot interpreted the transfer of heat as resulting from the movement of calorics going "down" from hot to cold sources. Just as water never goes back up by itself, calorics never flow by themselves from cold to hot. Just as in a perfect water-powered machine, the work done in a perfect steam engine would be enough to "pump" back the heat (calorics) from cold to hot. He found out that a particular kind of thermodynamic cycle, called now the Carnot cycle, would allow just that. However, even such an ideal machine could only transform into mechanical energy some fraction of the heat flowing down from the boiler to the condenser. Carnot did not invent the name "entropy", but the concept is really right there. As we shall see, entropy production in his theoretical machine is zero, which is why it has the highest possible efficiency. Notice that Carnot arrived at the right conclusion starting from a wrong concept of what heat is and well before the concepts of energy and entropy (which means transformation in Greek) were finally clarified by Rudolf Clausius, more than thirty years later (the final formulation of the two laws of thermodynamics dates from 1872).

So let us go through the description of thermodynamic cycles. We can think of a rudimentary steam engine as comprising a boiler producing the hot high pressure steam, a piston moving inside a cylinder and a wheel turning under the action of the piston (Fig. 4.1).

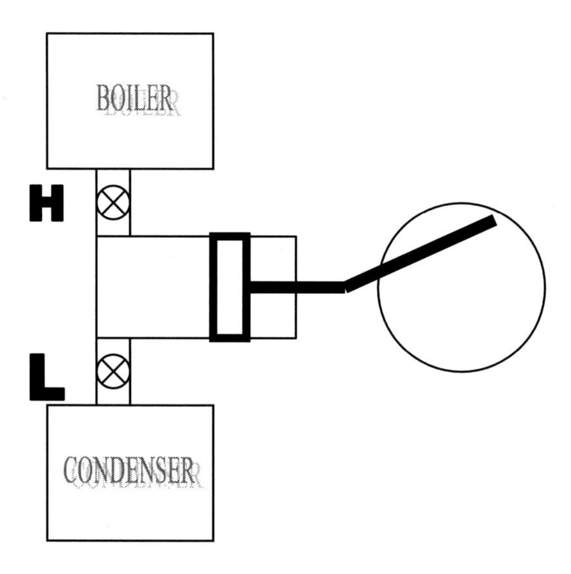

Fig. 4.1. Schematic representation of a rudimentary steam engine. Each cycle consists of the following 2 steps: (i) Valve **H** is open, steam is introduced inside the cylinder, the piston moves to the right which gets the wheel turning; (ii) When the piston gets to the end of the cylinder, valve **H** is closed and valve **L** is open. The piston moves back under the inertia of the wheel, steam is evacuated in the condenser. Then a new cycle can start.

The boiler is connected to the cylinder through a valve **H** (hot) that can be opened or closed and the steam in the cylinder can be released into the condenser through another valve **L** (low). Let us follow the evolution of the cycle on a pressure — volume diagram (Fig. 4.2). Here the pressure is that of the steam and the volume is that which it occupies. At the beginning of the cycle — point A — the valve **H** connecting the boiler to the cylinder is in the open position, valve **L** is closed. At this point the pressure applied by the steam to the piston is equal to P_A and the steam occupies a volume V_A. The piston gets moving, the wheel starts to turn. After a certain amount of steam has been admitted, valve **H** is closed. The piston keeps moving, the volume occupied by the steam expands and its pressure decreases. When the piston almost reaches the right end of the cylinder, we are at point B of the cycle. The pressure and volume are P_B and V_B. Then valve **L**, connecting the cylinder to the condenser is opened, and the steam is released. There is a sudden drop of pressure, the volume still expands a bit until the piston reaches the end of the cylinder. Point C is reached. The piston then moves back thanks to the inertia of the wheel

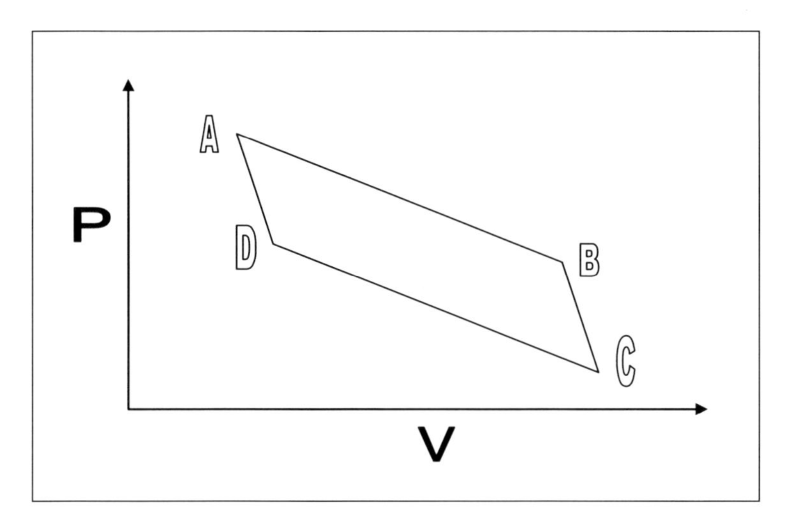

Fig. 4.2. A schematic thermodynamic cycle for a steam engine in the pressure-volume plane.

which keeps turning. The remaining steam is pushed out of the cylinder. Just before the piston is back to its initial position, at point D of the cycle, valve **L** is closed. Valve **H** is opened again, pressure goes back up while the piston completes its back motion. We are back at point A. A new cycle can start as fresh steam is introduced into the cylinder.

4.1.3. *Work performed in a thermodynamic cycle*

For the engineer the efficiency of the engine is equal to the amount of work accomplished during the completed cycle ABCD, W, divided by the amount of heat introduced into the cylinder from the boiler, Q_H. This efficiency, η, is called conversion efficiency, as it is related to the conversion of one form of energy into another.

$$\eta = \frac{W}{Q_H} \tag{4.1}$$

In order to calculate η we must first calculate W.

Let us consider one particular cycle for which this calculation is easy. As shown in Fig. 4.3, let us assume that the pressure is constant and equal to $P(H)$ along the segment AB, the volume constant along BC, the pressure constant and equal to $P(L)$ along CD and again the volume constant along DA. The force F exerted on the piston is equal to the pressure $P(H)$ times the area S of the piston:

$$F = P(H).S \tag{4.2}$$

The work performed along AB is the force times the displacement D of the piston:

$$W = F.D \tag{4.3}$$

Since the displacement D along AB times the area S of the piston is equal to the volume difference $[V(A) - V(B)]$, we have finally for the work performed along AB:

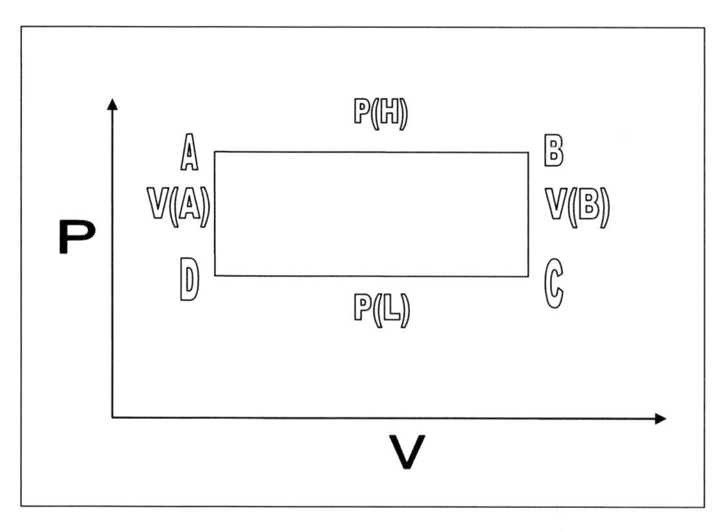

Fig. 4.3. Thermodnamic cycle allowing an easy calculation of the work done which is given by the area of the rectangle ABCD.

$$W_H = P(H) \,.\, [V(B) - V(A)] \tag{4.4}$$

Along BC the piston does not move thus no work is done. Along CD the work is:

$$W_L = P(L) \,.\, [V(A) - V(B)] \tag{4.5}$$

Again along DA no work is done. The total work done is:

$$W = W_H + W_L \tag{4.6}$$

$$W = [P(H) - P(L)] \,.\, [V(B) - V(A)] \tag{4.7}$$

which is equal to the area of the rectangle ABCD.

It can be shown that this result is in fact general, namely that the work accomplished during one cycle is always the area enclosed inside the four segments of the cycle.

4.1.4. *The Carnot cycle*

Different types of cycles have different conversion efficiencies. As said above, it was Carnot who introduced the idea of a cycle that has the highest possible conversion efficiency.

At this point we are considering an idealized engine where no energy is lost through friction between the piston and the cylinder and other kinds of losses due to imperfections of the machine are also neglected.

The diagram shown in Fig. 4.3 was helpful in showing how the work done can be calculated, but it is in fact a complicated cycle because there are temperature changes and heat exchanges along each leg of the cycle. It is helpful to consider a cycle where along legs AB and CD the temperature is constant — such legs are called isotherms — and where along legs BC and DA there is no heat exchange — such legs are called adiabatic. This is the Carnot cycle, shown in Fig. 4.4.

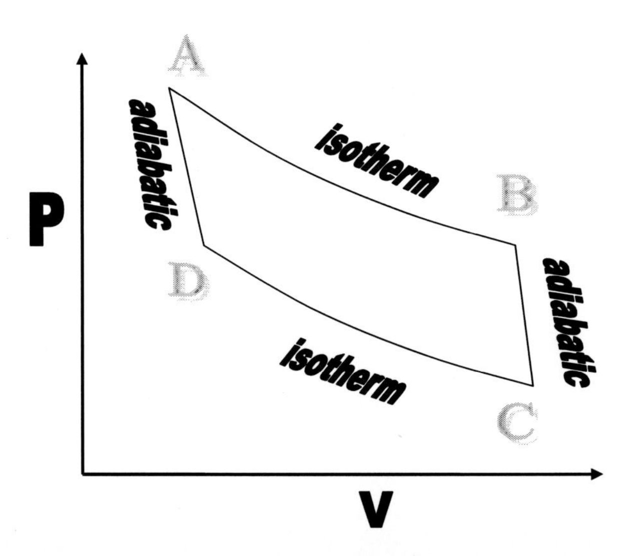

Fig. 4.4. Schematic representation of a Carnot cycle. Along the isotherm AB, heat is introduced at a constant temperature. Along the isotherm CD, heat is evacuated at (another) constant temperature. Along BC and DA, no heat is exchanged.

The calculation of the conversion efficiency of the Carnot cycle is very easy, as we now show.

Along AB, an amount of heat Q_H is introduced into the cylinder. Along BC, no heat exchange occurs. Along CD, an amount of heat Q_L is removed from the cylinder. Again along DA, no heat is exchanged. The total heat input around a cycle is therefore $(Q_H - Q_L)$. According to the first law of thermodynamics, this must be equal to the work performed since energy is conserved:

$$W = (Q_H - Q_L) \tag{4.8}$$

As mentioned above, the conversion efficiency is defined as the work produced divided by the heat Q_H introduced into the cylinder from the boiler:

$$\eta = \frac{Q_H - Q_L}{Q_H} \tag{4.9}$$

or:

$$\eta = 1 - \frac{Q_L}{Q_H} \tag{4.10}$$

Thus the conversion efficiency is always lower than unity, even in our idealized machine. Its value depends on the ratio of the heat removed from the cylinder into the condenser, to that injected from the boiler into the cylinder. Let us assume with Carnot that this ratio is equal to the ratio of the temperature of the lower isotherm, T_L (assumed to be equal to the temperature of the condenser), to that of the upper isotherm, T_H (assumed to be equal to the temperature of the boiler):

$$\frac{Q_L}{Q_H} = \frac{T_L}{T_H} \tag{4.11}$$

Then the conversion efficiency is simply:

$$\eta = 1 - \frac{T_L}{T_H} \tag{4.12}$$

This was the main result obtained by Sadi Carnot. The conversion efficiency of the best steam engine can never exceed (or even reach) the value given by Eq. (4.12), no matter how smart engineers will be. The temperature of the boiler must be as high as possible, and the temperature of the condenser as low as low as possible — both are equally important. If the temperature of the condenser could be equal to the absolute zero of temperature, the conversion efficiency would be equal to unity.

4.1.5. *Entropy change: introducing the second law of thermodynamics*

In order to simplify matters, we have given Q_H and Q_L positive signs, but this was not quite logical since Q_H is heat *added* to the system and Q_L is heat *removed* from the system. They should thus have been given opposite signs. The convention is to give the positive sign to heat added to the system, and the negative sign to heat removed from it. Equation (4.11) should then be written as:

$$-\frac{Q_L}{T_L} = \frac{Q_H}{T_H} \tag{4.13}$$

In this way we have introduced a new quantity, the ratio of the heat input (positive or negative) along a path to the temperature on that path (which is here constant since paths AB and CD where heat is injected and removed are isotherms). This is the definition of the entropy change along that path:

$$\Delta S = \frac{\Delta Q}{T} \tag{4.14}$$

where the symbol Δ stands for change in the quantity that follows. If heat has been put in, as it has along the path AB, entropy has increased. If heat has been removed, as it has along the path CD, entropy has decreased. The cycle invented by Carnot is such that the entropy increase along the path AB is exactly compensated by the decrease along the path CD, see Eq. (4.13). Along the paths BC and DA, there is no entropy change since no heat is put in or removed. Again, remember that we are considering an ideal machine. The reader could for instance rightly object that since at point B of the cycle valve **L** is open there must be some heat exchange along the leg BC. In any real machine, there will be some. The Carnot cycle is a theoretical one whose merit is to show that even in an ideal machine the conversion efficiency is limited for a very fundamental reason.

The Carnot cycle is such that *the total entropy change around the cycle is zero.* This is the reason why this cycle has the highest possible efficiency. Note that even though the efficiency of that best possible cycle is less than 100%, nothing has been lost. To convince ourselves that this is so, consider that we go around the cycle counter-clockwise: we put in work (which means that some other machine will have the wheel turning in the opposite direction), and by doing so we remove heat from the condenser and put it back into the boiler. Back to square one. The operation of this ideal machine is reversible, we have a case of perpetual mobile. Just as in the perfect water-powered machine considered by Sadi Carnot father, Lazare Carnot.

The Carnot cycle is an illustration of the second law of thermodynamics which states that the entropy in a closed system — here the combination of boiler, engine and condenser — can never decrease. In an ideal set up, it could remain constant, but in fact in any reality it will go up.

The devil is in the details. For instance, there will necessarily be some friction between the piston and the walls of the cylinder. Some of the mechanical work produced by the movement of the piston will be transformed into heat, and the machine will be less efficient. Going around the cycle the excess heat produced by friction will destroy the entropy balance of the ideal Carnot cycle: there will be a net *entropy increase.* The reduced efficiency and the increased entropy go hand in hand. The machine is now *irreversible*. Something has been lost that can never be recovered. This loss is best expressed by the net entropy increase.

Now suppose that you have fixed the amount of work that needs to be done. Because the machine is less efficient, more heat energy will have to be put in to obtain it. To obtain the same work from the machine, the irreversibility (or increase in entropy) needs to be compensated by an additional energy input.

4.1.6. *Energy, entropy and free energy*

What we have learned up to now is that while energy is conserved, entropy is not. Changes in entropy really determine how much work we can get done with a given energy input, described in the above example — as heat from the boiler. Since the amount of work that we can get done is really what interests us, we want to introduce a quantity that is directly related to it. This quantity is the free energy, named F. It is constructed in such a way that its *variations* are equal to the work accomplished (or received). It is defined as:

$$F = U - TS \tag{4.15}$$

where $U = W + Q$ is the total energy. For instance, let us see by how much the free energy varies around a Carnot cycle. Going from point A to point B, the change in F is:

$$\Delta F(A,B) = W(A,B) + Q_H - T\frac{Q_H}{T} = W(A,B) \tag{4.16}$$

where we have used the definition Eq. (4.14) of the entropy change. Going from point B to point C, there is no heat input since we have selected for that part of the cycle an adiabatic line, therefore:

$$\Delta F(B,C) = W(B,C) \tag{4.17}$$

Going around the full cycle:

$$\Delta F(A,B,C,D) = W(A,B,C,D)$$

In this ideal reversible machine the loss of free energy of the system consisting of the boiler, piston and cylinder, and condenser, is exactly balanced by the work done against the external world by the moving piston. In a non ideal, irreversible machine, additional heat generated by friction as we have discussed above will produce a net increase of the entropy S and therefore a decrease of the free energy larger than the work done. While energy is always conserved, free energy is not. Irreversible effects will continuously decrease the free energy, or energy available to do work.

As we will next discuss, the concept of free energy is also useful in situations where the entropy increase is due to such phenomena as diffusion, of which we have given a few examples in Chapter 1. In order to make this connection, we need to have a look at the microscopic aspects of entropy changes.

4.2. Entropy at the molecular level

To start our discussion, let us take the example of a hot bath and a cold bath weakly connected together. By weak connection we mean here that there is a small "heat leak" between the two baths. Through this leak, some heat will flow from the hot bath to the cold one.

If, in parallel with the weak heat leak, we put an engine that functions between the hot bath and the cold bath (like the steam engine functions between the boiler and the condenser), this spontaneous flow through the heat leak will reduce the amount of work that the machine will have performed until the two baths are at the same temperature. The total work produced by the machine will be insufficient to take back the same amount of heat as was transferred from one bath to the other during the time the machine produced work.

We can immediately cast this story in terms of free energy if we accept that *heat flows spontaneously from the hot bath to the cold one, because this flow increases the entropy of the system.* As we can see from the equation defining the free energy, this increase of entropy leads to a decrease of the free energy, which in turn reduces the amount of work that can be produced.

In Chapter 1, we used the example of the household to make the point that an energy input from the outside is necessary to prevent the increase of disorder. We are now in a position to back up this statement by a precise physical example. Suppose that the "order" to be kept is the difference between hot and cold regions. Irreversible processes are equivalent to a small heat leak between these regions, which continuously increases the entropy of the system. In order to keep the hot region hot and the cold one cold, we can install a heat pump (based on a Carnot cycle working in reverse) to continuously pump back heat from the cold to the hot region. The work done by this reversed engine must be performed with the help of an outside source of energy. This is exactly how a refrigerator is kept cold in a warmer room.

All of this is now clear, but why does heat flowing from the hot bath to the cold one through the heat leak induce an increase of entropy? What happens at the microscopic level when hot and cold are mixed? And in the first place, what is hot and what is cold? While heat and temperature are macroscopic concepts, they must find their origin in the motion of the molecules that constitute the liquid (or gas, or solid) under consideration. Contrary to what Lavoisier thought, heat is not carried by "caloric" particles.

Indeed, the development of the notion of entropy and its microscopic interpretation are intimately linked to the existence of atoms and molecules, as we have already mentioned in Chapter 1. Today their existence looks to us as an evidence, but in fact its scientific proof was only given relatively recently by Boltzman and is intimately related to the topic that we are discussing here. In short, the notion of temperature in a gas is indeed related to the motion of the molecules that constitute it, more exactly, to their kinetic energy. This energy is, on average, higher in a higher temperature gas (or liquid) than in a lower temperature one — on average, molecules move faster in a hot bath than in a cold one and have therefore a higher kinetic energy. In a system consisting of a hot bath and a cold one weakly linked together, a high kinetic energy molecule is on average more likely to be in the "hot" part than in the "cold" one, and vice versa. By measuring the kinetic energy of a molecule, we can with some degree of probability predict where it is. It is like distinguishing between white and black objects. If these objects are grouped separately in space, by just looking at their color we know where they are. If they are mixed randomly, the color does not tell us anything concerning their location. In the process of mixing hot and cold, or black and white, some information has been lost. This loss of information is one way to interpret the "natural" tendency to mix (here in space) different species. The loss of information corresponds to an increase of entropy.

Let us take a different example where heat is not involved. Consider a row of spheres, all of them black except for a white one which is located at one end of the row, say the left end to be specific. Now assume that there is a small probability that neighboring spheres can interchange their positions. After some time, the white sphere could be found anywhere, even on the right end side of the row. What we have done is give the white sphere the possibility to occupy any position on the row. The original order (the white sphere is at the left end of the row) has been destroyed, entropy has increased. What Boltzman showed was that the increase in entropy that occurs when we give the black sphere the possibility to occupy all positions on the row is equal to the logarithm of the number of available positions. At any finite temperature, the state of maximum entropy (the black sphere can be anywhere) is the preferred one because it gives the lowest free energy.

The above example is a simple model for very important phenomena such as pollution. Let us say that the white sphere is an impurity. In the beginning it was located at one end of the row, the rest of the row was "clean". As time goes on, the white sphere can be found anywhere, the row of black spheres has been "polluted". This is exactly how our cities are being polluted by molecules and particles emitted by cars, for instance. They come out of the exhaust, and after a while can be found anywhere. They mix with the molecules of “clean air”, nitrogen and oxygen for the most part. This process is unavoidable because it increases the entropy and decreases the free energy. It cannot be fought easily, once the impurities are in the open. Filtering the air of our cities is not a practical proposition. What we can do — and in fact what people do in highly polluted cities — is to breathe through a filter. Not surprisingly, this requires work.

4.3. Energy needs and man generated entropy

Equipped with this better understanding of the notions of entropy and free energy we can now go back to our energy needs.

First, let us consider our own essential personal needs, namely food energy. We do consume about 100 W and this power does not increase much when we exercise moderate physical activity. So what is this power needed for? The answer is now obvious: it is needed to compensate the entropy increase that occurs continuously in our body as a result of complex irreversible processes that are linked to mechanisms necessary to keep it alive. We have no control over these processes. If we fast, the body will feed itself on its reserves until they are exhausted and then death will ensue. There is no way we can train our body to consume permanently substantially less than these 100 W. There is this old story of the camel that almost got used to go around without eating when it unfortunately died.

We can compare our body to an idling machine which consumes substantial power in the “stand by” position (like a computer or a car blocked in a traffic jam). The main difference is that the body-machine cannot be turned off and on again. We live or we don’t. The very fact that we consume power even when we do not do any work is the result of irreversible processes that produce an entropy increase, which must be compensated for by an external energy input, here food. These processes are necessary to sustain life. We do not need to understand them in detail to draw the conclusion that they must be compensated for by an energy input. We can imagine that the amount of power we consume is the result of a long Darwinian evolution that has optimized it. In fact, if we remark that the power we consume is about the same as that needed for a personal computer, we can only marvel at the perfection of the human being from a thermodynamic standpoint. We do so much more than a PC with the most advanced circuitry. This comparison is interesting in its own right. The PC does not perform any physical work and the only reason it needs power is to compensate for the irreversible processes that occur while it treats information. We have already noted that there exists a relation between information and entropy.

We can now understand better our energy needs at the society level, if we compare society to some gigantic organism that needs power just to keep going, even if does not do any physical work. We say that we go "to work" everyday, but in fact most of us do not produce any work at all in the usual physical sense. It is really the organization of society that requires that we go to work. If everybody will stop working, society will collapse very quickly, and we can think of such a collapse in terms of an entropy increase. We can describe our activities in terms of "cycles" during which little or no physical work is accomplished, but during which power is consumed. For instance, we use public or private transportation to go to and come back from work, and in doing so "energy" is used. At work, we consume electricity and heat. If we work in an office, as many of us do today, no physical work is done at all. All of this power is consumed in order that we accomplish our tasks and that society keeps staying "alive". Our activities are the life sustaining processes that keep society functioning properly. They are needed to avoid the entropy increase that would occur otherwise, and would result in the "death" of society. Just as food is needed to keep us alive, an energy input is necessary to facilitate these activities.

The problem is that the power that is consumed in modern societies per person just to keep society going is huge compared to our own personal basic energy needs for food. As we have seen in Chapter 3, these needs are of the order of 10,000 W (more in Northern America, less in Europe), one hundred times larger than our food energy needs. This power is so large that if it would be extended the world over, insurmountable problems would ensue. Modern societies are in fact pretty bad thermodynamic machines, they generate a lot of entropy that must be compensated for by a huge energy input. This energy is generated by machines (such as the steam engine connected to an electricity generator) which themselves generate additional entropy, because irreversible processes occur in all the machinery we use. More energy is needed to compensate for it, and so on. For instance, large scale fossil fuel burning results in the release of gases (increases entropy) which according to many models produces average ("global") warming, which will have to be compensated for by an increased use of air-conditioning, which itself augments energy needs and so on. This can be a diverging process, which is precisely the entropy crisis.

This analysis is useful because it underlines that a solution to the "energy crisis", to "global warming" and other forms of damage to the environment cannot be found at the individual level. As already noted, we, as individuals can only "save energy" (release less entropy) to a limited extent. It is the way in which societies themselves function and the machinery they use that will have to be modified to substantially reduce the rate at which entropy is produced. In a Darwinian sense, this reduction is a necessity for the survival of modern society.

Chapter 5

Climate Change: What We Know and What We Don't

The term climate change refers to a mix of variables including temperatures on earth and in the oceans, ice cover at the poles and elsewhere, amount and nature (rain or snow) of precipitation, and variation of these quantities depending on time and location. Causes of climate change are numerous. They include:

(1) astronomical effects such as the variability of solar radiation, changes in the eccentricity of the orbit of the earth around the sun and of the inclination of its axis of rotation, possibly motion of the solar system itself through the galaxy, impact of large scale meteorites.
(2) geological effects such as tectonic plate motions, the distribution of emerged continents on the globe and volcanic activity.
(3) the effect of living organisms on the composition of the atmosphere particularly greenhouse gases (primarily the concentration of carbon dioxide).

The impact of these different causes for climate change is itself dependent on the interaction and exchanges between the oceans and the atmosphere.

The multiplicity of causes for climate change and of the above interactions makes it risky to attribute any observed change to a single cause.

In Chapter 2 we briefly described the history of the biosphere over a time scale ranging from the billion to the million years. This is the relevant scale for the development of life on earth. Here we shall cover another factor of one thousand from the million to the thousand years, which is the scale relevant to the development of mankind. In the end we shall focus our attention on the last 10,000 years, during which organized forms of societies have evolved. On the time scale relevant for this chapter there have been no important geological events affecting the climate, but as we shall see both astronomical effects and their interaction with the biosphere have been important.

5.1. Time scale and temperature scale

Climate change is, and should be, of major concern to us since our civilization has developed thanks to a rapid temperature elevation that occurred 10,000 to 20,000 years ago. Climate has been stable ever since and we naturally wish it to remain as it has been and still is today. We shall put this recent temperature elevation in context shortly but before we get there, it is important to note the scale of the temperature variation we are interested in.

The average temperature at the surface of our planet is above the temperature of melting ice, the zero of the Celsius scale. This average temperature, which is the result of a balance between the incoming solar radiation and the outgoing radiation from earth into space, has tremendous consequences. Since the temperature at the earth's surface varies by several tens of degrees from the poles to the equator, the former are below freezing and the later above it. Changes in the average temperature by only a few degrees Celsius will move the border between parts of the earth's surface that are "below freezing" and those that are "above freezing". A change of the average temperature by 10 degrees Celsius takes the earth in and out of a period where glaciers cover a substantial fraction of the northern hemisphere, where most of the emerged lands are found. So we must pay attention to small changes.

5.1.1. *The earth's temperature over the last few hundred thousand years*

Much of our discussion will involve the occurrence of ice ages. Let us first recall that an ice age is defined as a period of time where both poles are permanently covered by ice. In general ice ages occur when continents extend between the two poles and block the free circulation of ocean currents around the equator, as discussed in Chapter 3. As we have seen in Chapter 4, heat flows from hot to cold regions so as to increase the entropy of the system and decrease its free energy. We should then expect warm water to move from the equatorial regions to the poles, and this would prevent ice formation. But this flow must then come back to the equator. Continental masses stretching between the two poles largely block these circular currents, which results in large ocean temperature differences between equatorial and polar waters. It is thus believed that the repartition of land masses has a large effect on the climate quite independently from other factors such as the concentration of green house gases in the atmosphere.

Since both poles are (still) covered by ice, we are in an ice age. This has been the case for the last 3 million years. However, within that period of time there have been important variations of the temperature and of the total ice volume on earth. Figure 5.1 shows how temperatures have varied over the last 5 million years as calculated from variations in the ^{18}O isotope content measured in sediment cores in Antarctica.

The first thing to notice on this graph is that the average temperature has decreased over the 3 million year period by about 8 degrees Celsius. This is the scale of temperature variation we are

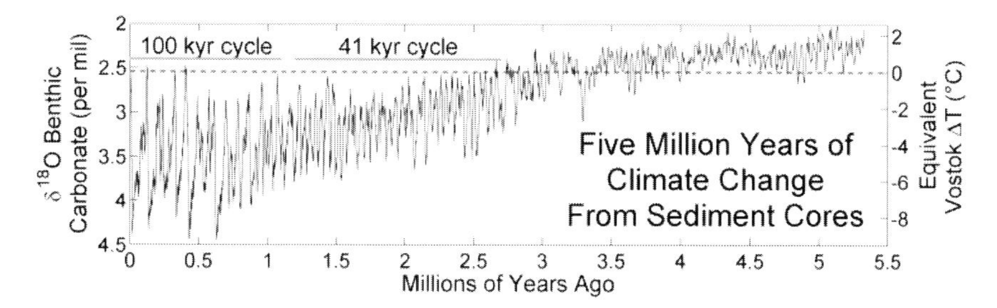

Fig. 5.1. The temperature at an Antarctic location called Vostok, over the last five million years, has decreased on the average. But there have been large fluctuations, particularly during the last million years, with amplitudes of the same order as the average temperature decrease (after L.E. Lisiecki and M.E. Raymo, Paleoceanography **20**, PA 1003 (2005)).

interested in. Temperature variations of less than one degree can be considered as small, variations of several degrees as large. It is important to keep this order of magnitude in mind for a meaningful discussion of current temperature variations.

The second thing to notice on Fig. 5.1 is that the overall temperature decrease is not smooth but shows large fluctuations. One can see that the periodicity of these fluctuations is not constant. They occur on a time scale of the order of 100,000 years during the last million years, and 41,000 years at earlier times. During the last million years the amplitude of these fluctuations, of the order of 8 to 10 degrees, is as large as the average overall decrease over the last 3 million years.

Figure 5.2 shows the variation of ice volume together with temperatures variations from two locations in Antarctica over the last 450,000 years, giving a closer view of these large fluctuations. A periodicity of about 100,000 years is clearly visible.

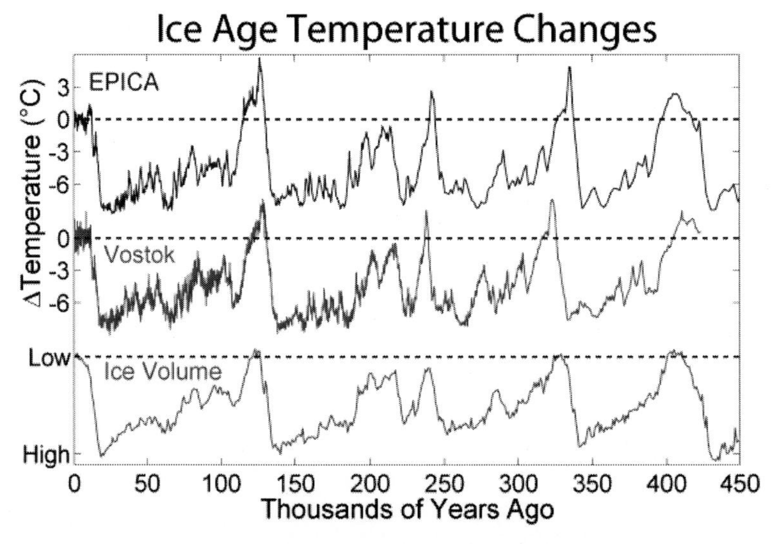

Fig. 5.2. Temperature variations at two Antarctic locations (two upper curves) and variations of ice volume (lower curve) during the last 450,000 years. Note the 100,000 years period, and the saw tooth shape of the ice volume variations (compendium from Wikipedia Ice Ages).

On the average, every 100,000 years there is a rapid decrease of the ice volume (increase of the temperature) followed by a slower increase (glaciation). The temperature rise has the form of a spike that lasts for about 10,000 years or less. These periods are called interglacial. The slower increase in the ice volume makes sense — it takes time for large volumes of ice to form.

Another important piece of data that comes from Antarctica ice cores measurements at Vostok concerns variations in greenhouse gases concentrations (carbon dioxide CO_2 and Methane CH_4). Figure 5.3 shows from top to bottom the variations of the temperature (a), of the CO_2 concentrations (b), of the CH_4 concentration (c) and two additional pieces of data, variations in Oxygen isotope ^{18}O in water (d) and the calculated insolation (bottom curve e).

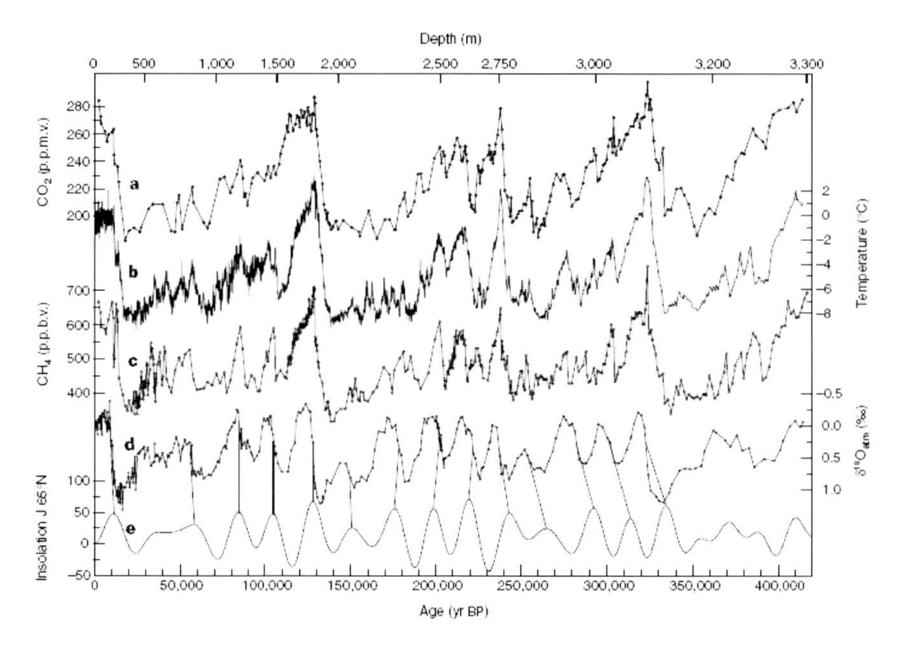

Fig. 5.3. Variations of (a) CO_2 concentration, (b) temperature, (c) CH4 concentration, (d) atmospheric content of the ^{18}O isotope in the atmosphere and (e) calculated insolation from astronomical variations. After J.R. Petit et al., Nature **399**, 429 (1999).

One central point that comes out from Fig. 5.3 is the perfect correlation between curves **a**, **b** and **c**, namely between temperature and greenhouse gases concentration variations. This holds not only for the main periodicity of 100,000 years, but even for detailed variations during interglacial and glacial periods. Starting from the right hand side one sees a broad interglacial peak, the two following ones are narrow and very spiky, and then comes a somewhat broader one.

To the left of the graph we can see the data for our current interglacial period. A rapid rise in greenhouse gases concentration and temperature occurred about 20,000 years ago, and then stabilized about 10,000 years ago (more exactly 12,000 years ago) at a level about 8 degrees Celsius above that of the previous glacial period.

This long period of temperature stability is unprecedented in previous interglacial periods. Periods of temperature stability during previous interglacial periods lasted more like 4,000 years.

Another important point to notice in Fig. 5.3 is the levels of CO_2 during glacial and interglacial periods. They have replicated fairly consistently around 180 ppm and 280 ppm respectively. The highest concentration, close to 300 ppm, was recorded for the interglacial period of 300,000 years ago. These are the reference levels for today's CO_2 concentration which is 380 ppm, certainly higher than was ever recorded at any time for the last 500,000 years or so.

Since the current interglacial period started about 20,000 years ago, one might expect a strong decrease of the temperature (end of the interglacial period) on the 10,000 years time scale, if the pattern observed in the last million year period repeats itself. This temperature decrease should be of about 8 degrees Celsius. It would be a drastic change in our climate, with a return of major glaciers down to the latitudes (New York, Northern Europe for instance) where they were 20,000 years ago. Since 10,000 years is the time scale of our civilization, the major threat to its survival is the end of that period. This may sound paradoxical or provocative at a time when attention is centered on the dangers incurred because of global warming. But the thought that global cooling rather than global warming constitutes the major threat is not unreasonable if our civilization is to last say another several 10,000 years. It certainly deserves attention.

5.1.2. *How well understood is the periodicity of interglacial periods*

The above view assumes that the duration of the current interglacial period will be similar to that of previous ones. This is a reasonable hypothesis, but one that needs to be backed up by an understanding of the origin of the observed periodicity. Possible mechanisms have been briefly mentioned in Chapter 2. In view of the great importance of this issue in the current context of feared climate changes, we now review them in more detail.

The periodicity of temperature changes is so clear that it leads one to think that it must be of astronomical origin, namely that they are primarily due to periodic changes in the amount of solar radiation received at the earth's surface. This point of view is widely accepted today. As said above, we can eliminate from our considerations geological factors that have been decisive in the past in triggering long lasting ice ages, such as movements of continents, which are irrelevant on the time scale we are now considering.

So let us see how astronomical factors could trigger alternating periods of ice cover extension and recession. The total amount of radiation received by earth depends on its distance from the sun, and its seasonal variations on the angle that the axis of rotation of the earth (called *tilt*) makes with the plane of its orbit around the sun. If earth's orbit around the sun was a perfect circle, the total amount of radiation would remain constant around the year, but this orbit is in fact an ellipse. Since the amount of radiation received varies as the inverse of the distance squared, earth receives more radiation when it gets closer to the sun, and vice-versa. This change has a periodicity of one year, the time it takes earth to complete one revolution around the sun.

It turns out that both the elliptical character of earth's orbit — called the eccentricity e — and the *tilt* of its axis of rotation are not constant but vary in time. The eccentricity varies with a main period of 400,000 years, its amplitude being modulated with a 100,000 years period. As for the tilt, it varies with a periodicity of 41,000 years. One can expect these periodicities to show up in the average temperature of the earth and corresponding extent of ice cover.

Indeed, as can be seen from Fig. 5.1, the 100,000 year and the 41,000 year periods are clearly observed. But there are complications. For instance, the 400,000 year period (which should be the dominant one) is not seen.

5.1.3. *The Milankovitch cycles*

Milankovitch considered, in addition to the variation of the *eccentricity* and of the *tilt*, the *precession* of the axis of rotation of the earth around the perpendicular to the plane of the orbit, which has a period of about 20,000 years (clearly visible in curve (e) of Fig. 5.3). He also took into account that there is a strong asymmetry between land masses in the two hemispheres, most of them being located in the northern one. Actually glaciation periods involving the advance and retreat of glaciers refer mostly to that hemisphere.

The sum of these variations is known as the Milankovitch cycles.

Let us for instance consider the effect of a variation of the *tilt* angle of the axis of rotation of the earth on its orbit. For smaller tilt angles, seasons are less pronounced, since insolation varies less around the year. One can argue that small tilt angles will favor glaciation because the larger insolation in the winter will favor larger evaporation from the oceans and therefore larger snow precipitations, and the cooler summers at high latitude will lead to reduced summer melting of the snow-ice. As said above, this effect has a periodicity of 41,000 years, which has been clearly observed (Fig. 5.1), with glaciation periods associated with smaller tilt angles and glaciers retreat with larger ones as expected.

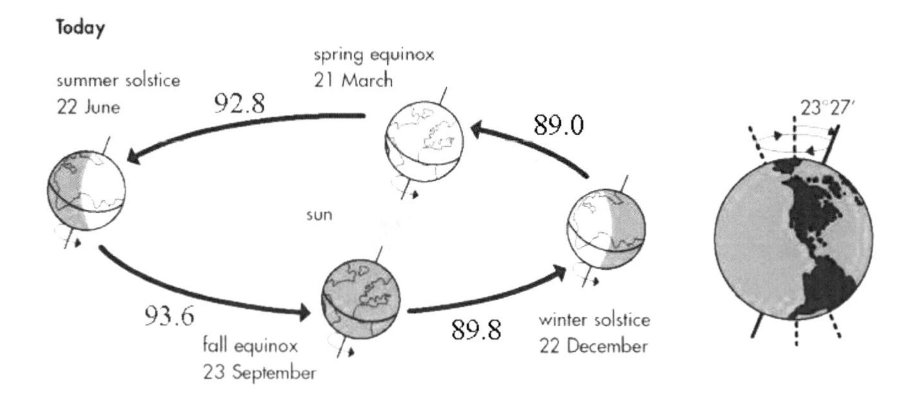

Fig. 5.4. Motions of the axis of rotation of the earth. Its inclination or *tilt* varies over a period of 41,000 years, and it precesses with a period of 21,000 years. This precession inverts the seasons, the summer solstice becomes the winter solstice after 10,500 years. The eccentricity of earth's orbit varies with a periodicity of 400,000 years, with a 100,000 years modulation (After Berger and Loutre, La Recherche **368**, 42 (2003)).

The *precession* of the axis of rotation of the earth, called *climatic precession,* combined with variations of the eccentricity, can also lead to glaciation periods. If the winter solstice falls when the earth is closest to the sun, and the summer solstice when it is furthest from it, there will be milder winters and cooler summers — a situation that favors glaciation as we have discussed above. And of course vice-versa. These two situations are realized alternatively. The effect is evidently most pronounced when the eccentricity is large, which enhances the amplitude of the 100,000 years periodicity.

5.1.4. *Problems with the Milankovitch cycles*

Several problems have been noted with the model of Milankovitch. The first one is that the calculated amplitude of the insolation variations is insufficient to explain the observed amplitudes of temperature variations. Another one is that according to calculations the leading periodicity should not be of 100,000 years as observed, but rather that due to the precession of the axis of rotation of the earth which has a periodicity of 21,000 years and produces the largest changes in insolation.

One would need an amplification mechanism if the measured temperatures changes were to be attributed to Milankovitch cycles. We have come across an amplification mechanism in Chapter 2 when we considered the effect of ice ages on carbon dioxide concentration in the atmosphere. It is related to the low albedo of snow-covered areas. If a decrease in insolation produces at first only a small increase in glaciers coverage, the enhanced light reflection triggers itself a further lowering of the temperature (more light being reflected, less is absorbed) and further extension of the glaciers, and so on. A further effect can then occur due to a reaction of the biosphere. Because earth is then less green, it captures less carbon dioxide from the atmosphere. Volcanic activity can then increase greenhouse gas concentration, leading to an increased temperature and end of the ice age. This argumentation is very qualitative, but shows that the interaction of astronomical insolation variations with reactions of the biosphere can lead to complex patterns of weather variation.

A fourth astronomical factor that was not known at the time of Milankovitch is the periodical variation of the angle that the earth's orbit makes with the averaged plane of all planets that orbit the sun. According to the Newtonian laws of mechanics, this averaged plane is an invariant, but each of the planets orbit can have its angle vary about it. It turns out that the periodicity of the variation of the earth's orbital plane is of 100,000 years. As the angle of earth's orbit varies it coincides periodically with that of small particles (may be residues of a planet) that also orbit the sun. When this coincidence occurs, the amount of insolation earth receives decreases as part of it is scattered away by the small particles. This hypothesis has been made rather recently and needs further work to ascertain if it can explain quantitatively the measured amplitude of temperature variations. It gives one the observed periodicities, but since other periodicities are observed as well (such as the 41,000 years period), it is not clear that periodical changes in the angle of earth's orbit is the main source of temperature variations.

Variation of the intensity of solar radiation is another factor to be taken into account. It is a well established fact from satellite observations that it follows an 11 year cycle, with typical relative amplitude of 1 part in 10,000. Accurate radiation measurements that have become available recently are well correlated with the activity of the Sun as measured by the number of sunspots. These have been observed for a long time, and offer one way to reconstruct how solar radiation has varied over several centuries. On that time scale, variations of solar radiation are the only astronomical factor to be taken into account. Changes can be fast, but their amplitude is small. They are believed to be irrelevant to ice ages.

5.1.5. *Towards a longer interglacial period?*

Because no consensus has yet been reached as to the exact mechanism of the periodicity of ice ages (and it is doubtful that there is a single one), it is difficult to make predictions as to their recurrence in the future. Berger and Loutre have proposed that the next ice age may be 50,000 years away or more, because we are at a point of the 400,000 years cycle where the eccentricity of the earth's orbit is at a minimum, which diminishes the impact of the precession effect as we have seen. At the same time, the tilt angle of the axis of rotation of the earth on its orbit is decreasing, which also diminishes insolation variations. All told, astronomical factors are such that at the moment the amount of radiation received by the northern hemisphere may be rather stable for a long period of time, which should lead to an anomalously long interglacial period.

To conclude this section on astronomical factors, it can be said that when they are considered in detail it is not clear that the current interglacial period should last for about 10,000 years as did previous ones (see Berger and Loutre for further reading).

5.2. The CO_2 cycles

An important element that is missing from current climatic models is an *ab initio* calculation of the variation of carbon dioxide in the atmosphere. This limits their ability to reproduce the CO_2 cycles as they have been recorded in detail for the last several hundred thousand years, see for instance, Figs. 5.2 and 5.3 which we have already commented upon, and *a fortiori* the validity of predictions one can make concerning future temperature variations based on astronomical factors alone.

It is recognized that greenhouse gases have a strong influence on the temperature at the surface of the earth, certainly on the same level as the astronomical factors discussed above, and this influence is today a major cause of concern. But current climatic simulations are not able to calculate self-consistently CO_2 levels based on astronomical factors and available models of the biosphere. These models can only reproduce past temperature and glaciation cycles if CO_2 data are put in "by hand". By the same token, they are not able to make predictions of CO_2 levels for the future, from which it follows that they cannot make predictions for temperature evolution. What they can do is use different scenarios for changes of the CO_2 atmospheric content, and on that basis make temperature predictions.

Figure 5.5 will illustrate this point. It allows a comparison between the variation of the eccentricity with that of CO_2 concentration. We have already noted from Fig. 5.3 that the latter perfectly correlates with temperature variations. The degree to which astronomical factors (eccentricity being supposedly the dominant one) can explain these variations can be inferred from Fig. 5.5.

The continuous line in Fig. 5.5 is the calculated variation of the eccentricity, and the more noisy line the measured CO_2 concentration at Vostok. Two observations can be made. The first one is that the periodicity of the eccentricity matches well that of CO_2 data, which is in favor of the astronomical model. But the second one is that there is no correlation between the amplitudes of

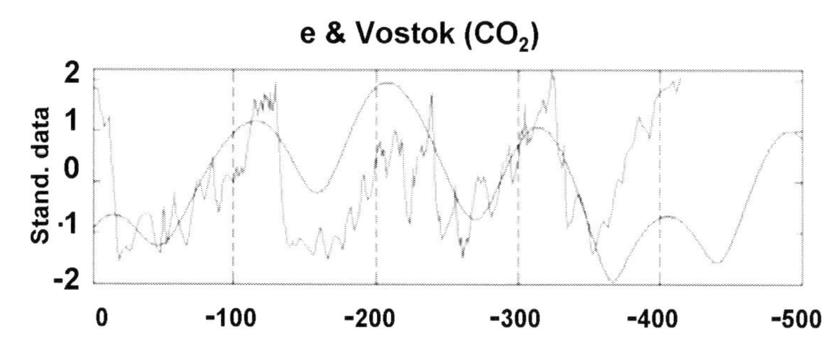

Fig. 5.5. Over the last 500,000 years the calculated eccentricity (e) of earth's orbit (continuous line) and recorded CO_2 concentrations vary both with a periodicity of 100,000 years but without correlation between amplitude variations. Eccentricity variations are strongly modulated (400,000 year period), but the amplitude of the CO_2 cycle is almost constant. Hence insolation changes (largely governed by eccentricity variations) are not sufficient to explain temperature variations during glacial-interglacial cycles (CO_2 concentrations are closely correlated with temperature variations, see Fig. 5.3). (After A. Berger, J.L. Melice and M.F. Loutre, Paleoceanography **20**, PA 4019 (2005).)

the oscillations of these two quantities. In the current interglacial period the amplitude of the eccentricity oscillation is extremely weak (as it was 400,000 years ago), but that of the CO_2 variation is quite the same as it was in the three previous ones.

Therefore eccentricity variations cannot fully explain the CO_2 and temperature cycles.

The same reservation can be made concerning the correlation between the variation of the total insolation, calculated taking into account not only variations of the eccentricity but also those of tilt and precession, and the temperature and CO_2 cycles. This can be seen Fig. 5.3 where the bottom line (curve **e**) is the calculated insolation. Its dominant period is that of the climatic precession, 20,000 years, while the dominant period for the CO_2 and temperature variations (curves **a** and **b**) is 100,000 years. Moreover, as can be seen at the left side of the diagram, insolation started to decrease about 10,000 years ago, which should have brought up the beginning of a temperature decrease, but CO_2 concentrations and temperatures have remained remarkably stable since then. A comparison between the various astronomical parameters — eccentricity, obliquity (tilt), climatic precession, total calculated insolation — and CO_2 concentrations — is shown Fig. 5.6. Curves shown cover the past as well as the future (for CO_2 it assumes repetition). No simple correlation appears between any of the astronomical parameters and CO_2 variations.

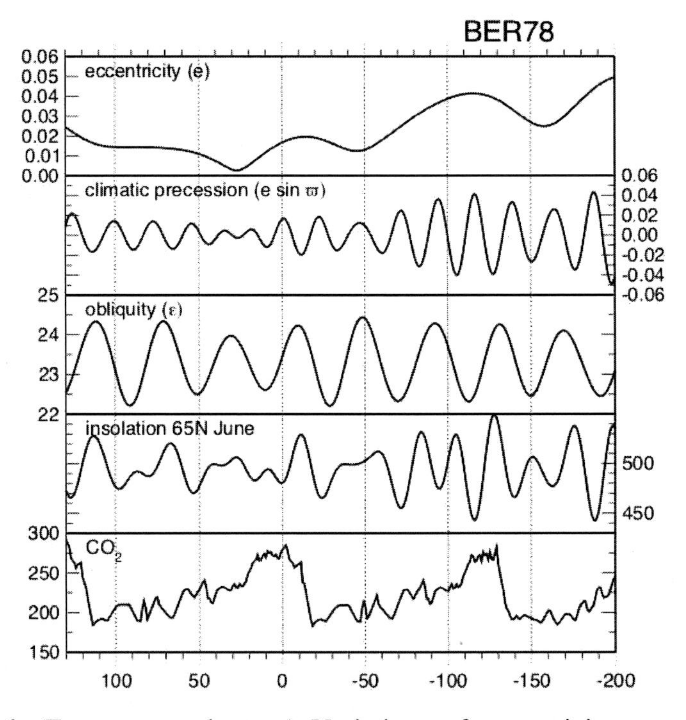

Fig. 5.6. (From top to bottom) Variations of eccentricity, precession, obliquity, summer solstice insolation at 65 degrees North, and CO_2 cycles from −200,000 years to +150,000 years. The future CO_2 cycle is a repetition of the last one. After A. Berger et al., Surveys in Geophysics **24**, 117 (2003).

Since CO_2 concentration and temperature variations do not correlate well with variations of the total insolation, or of any of the astronomical parameters, it is not obvious how variations of the latter can be the primary source of climate change. We really have no simple explanation for the perfect correlation between CO_2 and temperature variations observed during the last 500,000 years. Additionally, one can only marvel at the repetition of the CO_2 concentration cycles, oscillating exactly between the 180 and 280 ppm levels. How could such a complex machinery with so many parameters reproduce them at such a level of exactitude? We must acknowledge the limited extent of our understanding of climatic evolution. We can, however, accept the correlation between CO_2 and temperature variations as an empirical fact. This is what we shall do in the next section.

5.3. Anthropogenic temperature changes

We are now equipped with the proper background to discuss the burning issue of anthropogenic climate change. The question is what impact have activities of man on the climate and what dangers does this impact entail. There was for many years a raging controversy as to the very existence of such a link. This controversy is now largely over. We shall see that there are indeed some very good and simple reasons to accept the present consensus that such a link does exist. The question is now quantitative, not qualitative.

So let us first briefly summarize the climate background at this point of the history of the earth. For the last several hundred thousand years temperatures and CO_2 levels have followed well defined cycles, with CO_2 levels rising from 180 ppm to 280 ppm and the temperature by about 8 to 10 degrees during interglacial periods. About 20 thousand years ago such rapid rises occurred, which resulted in the retreat of glaciers that had occupied most of Northern Europe and Northern America for close to 100,000 years. Temperature and CO_2 levels reached their maximum about 15,000 years ago, and have remained remarkably constant since then (except very recently). Similar levels were reached during previous interglacial periods, but the length of time where levels have remained stable is unique to the current interglacial period.

This climate stability is clear if we focus our attention on the last 80,000 years of the Vostok record. Figure 5.7 shows the variations of Deuterium concentration (a proxy of temperature variations) in ice as a function of depth. A depth of 1550 meters corresponds to about 110,000 years. At a depth of 350 meters, or 25,000 years ago, a rapid temperature rise becomes apparent. It is completed at a depth of slightly less than 300 meters, or 20,000 years ago. The temperature rise was completed in about 5,000 years. This means that during this period the temperature rose roughly by a quarter of a degree every century. This is a figure worth remembering for the discussion that follows.

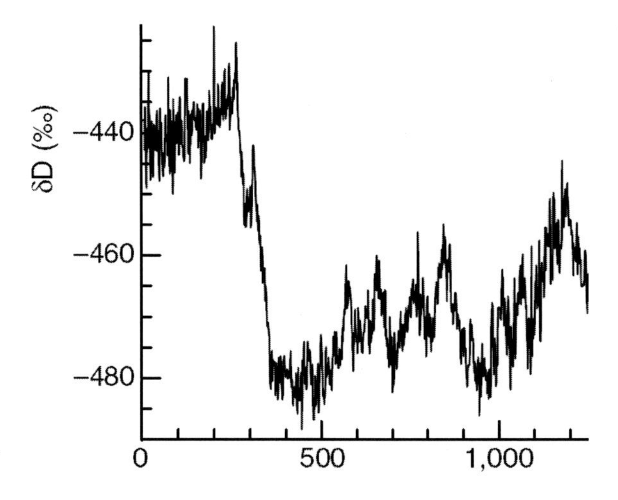

Fig. 5.7. Variation of Deuterium concentration in Vostok ice cores as a function of depth in meters. Deuterium concentration is a proxy for temperature variation. Ice cores taken from a depth of 500 meters were deposited about 35,000 years ago. On this enlarged graph the temperature stability achieved during the last 12,000 years (corresponding to depths less than 150 meters) is remarkable: it is unique compared to previous interglacial periods. After J.R. Petit et al., Nature **399,** 429 (1999).

5.3.1. *The CO_2 anthropogenic footprint*

Up to a few centuries ago, not much happened to the CO_2 concentration in the atmosphere. In fact, as recently as 1958, the CO_2 level was 315 ppm (Fig. 5.8), not much above the value of 280–300 reached during previous interglacial periods.

But, as can be seen Fig. 5.8, a change occured around 1975. Since then the CO_2 concentration has been rising much faster than it did before. From the mid-seventies it has been rising at a rate of about 15 ppm per decade. This rate should be compared to that typical of the beginning of interglacial periods, which is of about 100 ppm over a few thousand years, or about 1 ppm per decade. The current rate of increase of CO_2 concentration is therefore unprecedented. It is clear that this is not a natural phenomenon. The rate of increase is such that it must have an anthropogenic origin.

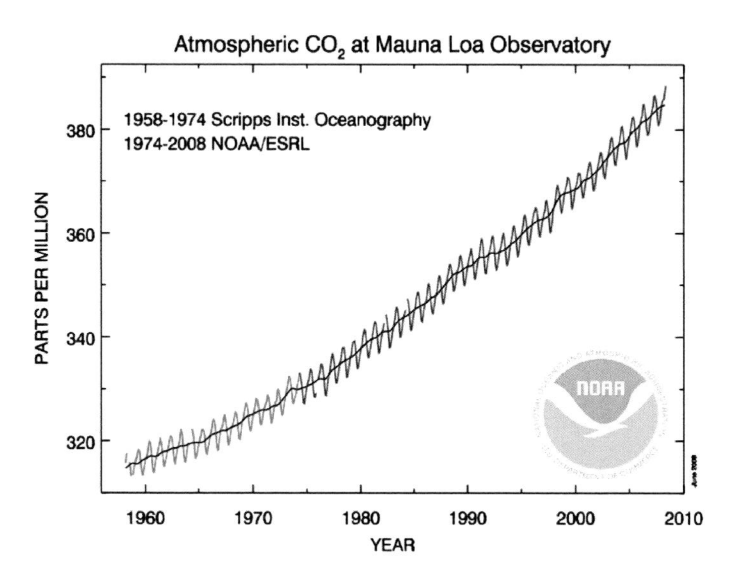

Fig. 5.8. Variation of CO_2 atmospheric content at the Mauna Loa observatory in Hawaii measured since 1958. At that time the CO_2 concentration was not much higher than maxima reached in previous interglacial periods (280 to 300 ppm).

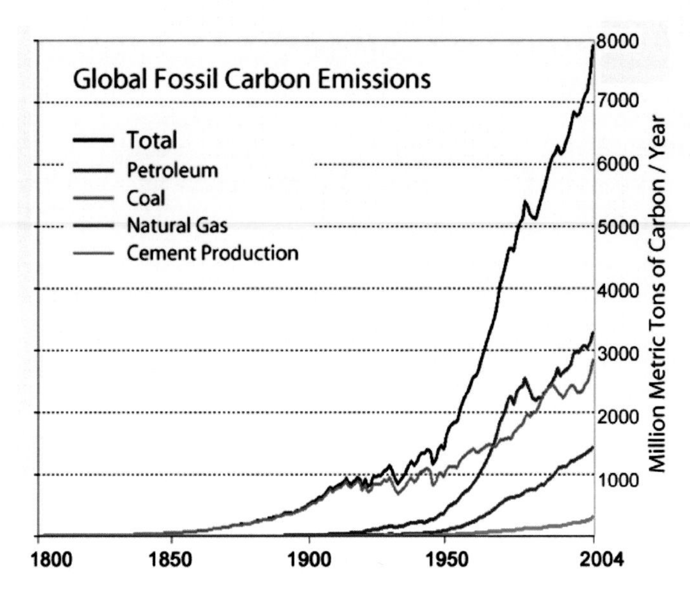

Fig. 5.9. Global Carbon emissions since the beginning of the industrial era. Note the fast rise of emissions from hydrocarbons (oil and natural gas) since 1950, and the maximum *rate* of increase reached around 1975. After G. Marland et al. in: Trends, a Compendium of Data on Global Change, Oak Ridge National Laboratory (2007).

The amount of CO_2 thrown into the atmosphere by the burning of fossil fuels is shown Fig. 5.9. It is often said that anthropogenic effects started at the beginning of the industrial era *circa* 1850, with the massive use of coal. Actually, as we have noted above, the rise of CO_2 concentration was modest until 1960, reaching a mere 15 to 30 ppm above the level typical of interglacial values added over about a century, or 1.5 to 3 ppm per decade. But another 70 ppm have been added in half a century.

The fast increase in CO_2 concentration after 1970 seen from Fig. 5.8 correlates with the fast rise in the use of oil, as seen from Fig. 5.9. Emissions of CO_2 from oil were negligible compared to those from coal until 1950. They reached comparable levels around 1970. Today CO_2 emissions from oil are as large as those from coal, the latter having risen steadily. The total rate of CO_2 emissions is to day 6 times higher than it was in 1950, and twice as high as it was in 1970. This fast rise is likely at the origin of the fast rate of increase of CO_2 in the atmosphere during that period, as seen Fig. 5.7.

It appears that as long as the rate of CO_2 *emissions* is below a certain level (maybe below the level reached around 1960 to 1970), the biosphere is able to react in such a way as to keep the CO_2 *concentration* in the atmosphere close to the 280 to 300 ppm typical of interglacial periods. Beyond that level, however, the biosphere is unable to hold in check the rise of CO_2 concentration.

A useful perpective of the evolution of CO_2 concentration over the last 1000 years is shown in Fig. 5.10.

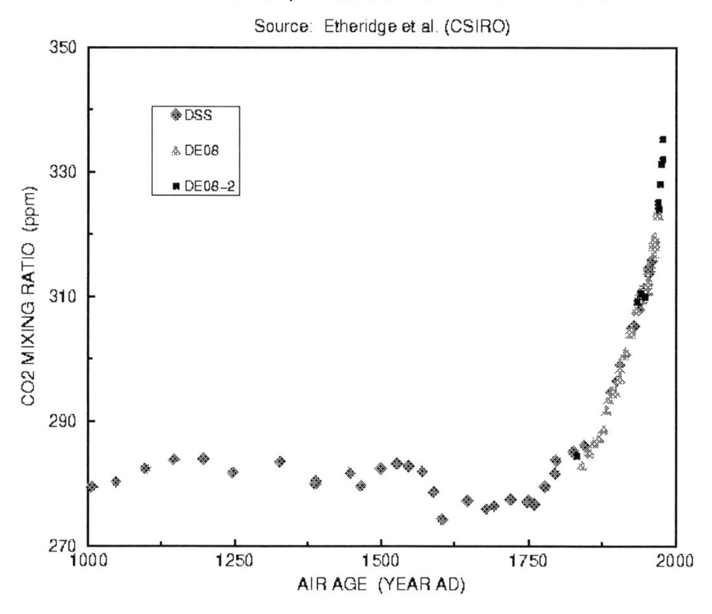

Fig. 5.10. Until about 1900 the atmospheric concentration of CO_2 remained close to the value typical of interglacial periods, 280 ppm, see Fig. 5.3. It is only in the second half of the 20th century that it has surpassed the highest value recorded in previous interglacial periods (300 ppm). After C.E. Ophard, Virtual Chembook, Elmhurst College (2003).

There were some ups and downs, but up to about 1900 this concentration remained in the vicinity of 280 ppm, equal to that recorded for previous interglacial periods. The rise since the last decades of the 20th century is so fast that on the time scale used in Fig. 5.10, it is practically vertical.

In the Vostok cycles the CO_2 increase of 100 ppm during interglacials (from 180 ppm to 280 ppm) is accompanied by a temperature rise of about 8 to 10 degrees Celsius. Taking this as a guide, we would have expected that the additional increase of 100 ppm seen since 1900, from 280 ppm to 380 ppm, should have been accompanied by a similar temperature rise of about 10 degrees. In fact, it has been of only 1 degree. This is surprising.

Maybe the temperature rise has so far been small because it takes time until the increased energy input due to a higher greenhouse gas concentration is translated into a temperature rise. Just as when we warm up food in the oven, there is a time lag between the calories input and the rise in temperature. The biosphere has a huge heat capacity and takes time to respond. Indeed, the temperature rise of 8 to 10 degrees at the beginning of interglacial periods took place over a period of several thousand years. Even if the CO_2 concentration does not increase more than it has so far, the temperature rise of about 1 degree seen since the beginning of the 20th century should be considered only as a beginning of a larger rise.

The surprising temperature stability in the face of the large rise in CO_2 concentration may also have a more subtle origin. Just as the CO_2 concentration in the atmosphere did not, for a while, follow emissions, it could be that the biosphere was able to prevent for a while the expected temperature rise through some complicated stabilization mechanism.

5.3.2. *The temperature rise in modern times*

It is now time to take a more detailed look at temperature records in modern times. The question is not anymore whether temperature has risen in modern times — it is now well documented that it has.

The recent report of the Intergovernmental Panel on Climate Change (IPCC) is very specific on that point, and we shall quote it extensively here below. The question, or more exactly the questions are rather:

(1) how much of the past temperature increase is anthropogenic and how much is not.
(2) how big a further increase is to be expected and on what time scale.
(3) what will be the consequences of this further rise?

5.3.2.1. *Evolution of the temperature since 1900: the start of anthropogenic effects*

It has often been said that the beginning of the industrial era, dated around the beginning of the 19th century, is the time when temperatures have started to rise beyond what natural causes would have produced. The last computer simulations compiled in the 2007 IPCC report show that this is not quite so. The data used in these simulations include insolation variations due to astronomical factors and solar activity, measured greenhouse gas concentrations and other factors such as aerosol concentration. The models include interaction between sea, land and atmosphere, and are able to reproduce past temperature and ice volume variations. Measured and calculated temperature variations fit quite well. They can be compared to temperature variations calculated without anthropogenic factors (without greenhouse gas emissions).

Results published in the IPCC report are reproduced Fig. 5.11. The temperature elevation since the beginning of the 20th century that appears in the graphs is of about one degree. This is the number that is usually quoted as the significant number that characterizes anthropogenic effects. However, this figure — one degree over one century — does not tell the most significant part of the story. A closer look at the graphs will reveal to the reader that there is no clear evidence for any anthropogenic effect until the mid 1970s.

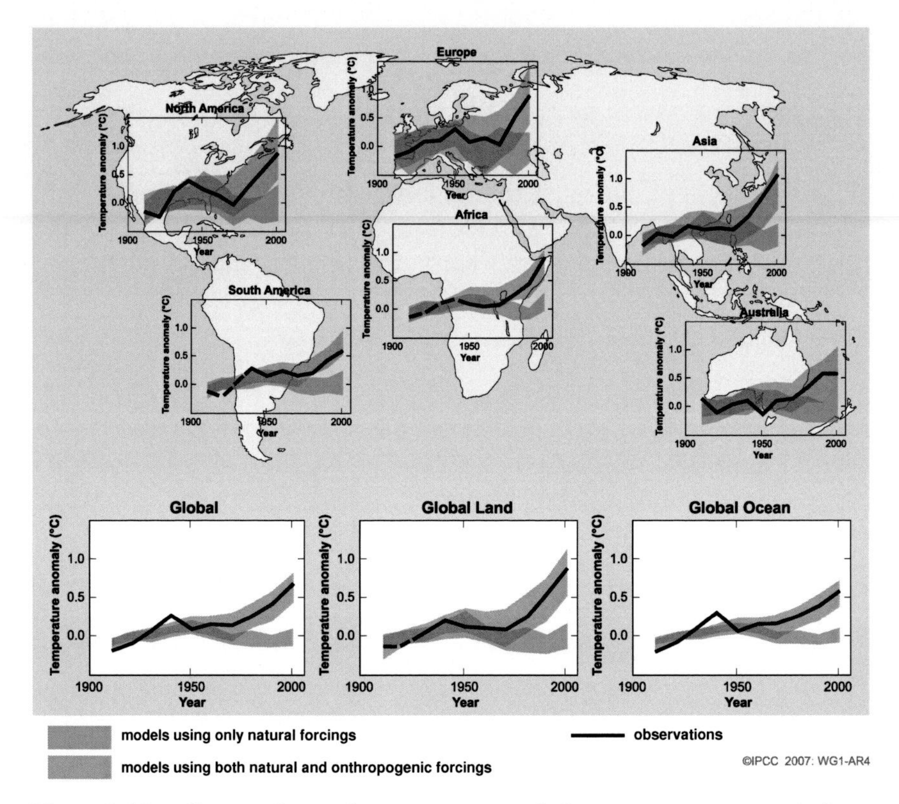

Fig. 5.11. Comparison between recorded temperature variations (continuous lines) and calculated ones with (pink bands) and without (blue bands) anthropogenic forcing, in the five continents (upper five graphs) and global, global land and global ocean (three lower graphs). Calculated temperature variations including anthropogenic forcing reproduce very well recorded temperatures over the past century. It is only since the last quarter of the 20th century that calculated temperatures with and without anthropogenic forcing have become clearly different (2007 International Panel on Climate Change, Summary for Policy Makers).

The graphs shown Fig. 5.11 present the variations of three temperatures. The dark continuous line represents the measured temperature change, the pink shaded area the calculated temperature variation including variations in insolation (natural forcing) and greenhouse gas emissions (anthropogenic forcing), and the blue shaded area the calculated temperature variation excluding anthropogenic forcing. The three lower graphs represent the global, global land and global ocean temperatures. In these three graphs, the three temperatures are indistinguishable up to about 1975, within the width of the calculated evolutions, which represents the range of a large number of different simulations. It is only since 1975 that calculated temperature changes which include anthropogenic forcing are in much better agreement with measured temperatures than simulations that exclude them.

A possibly important conclusion can be drawn by comparing Fig. 5.8, Fig. 5.9 and Fig. 5.11. Emissions prior to 1970 (mostly from coal combustion) did not result in a measurable temperature elevation. Beyond 1970, the rate of emissions grew faster, now dominated by oil and gas (Fig. 5.9). At the same time, the CO_2 atmospheric concentration started also to grow faster (Fig. 5.8). It is only then that an anthropogenic effect on the temperature clearly appears. This may mean that as long as the rate of greenhouse emissions is below a certain level, the biosphere reacts in such a way that the impact on the CO_2 concentration and on the temperature rise is small. Beyond that emission rate, the biosphere and the climate may become unstable.

Additionally, it could be that the larger amount of particulates (soot) emitted by coal combustion as compared to oil and gas did mitigate to a certain extent the greenhouse effect by reducing the amount of solar radiation reaching the surface of the earth.

Figure 5.12 shows a blow-up of the three temperature evolutions on land. This graph shows clearly that up to about 1970 to 1980, there is no clear indication of any anthropogenic effect on the temperature. But beyond 1980 there is a startling departure of the measured and calculated temperatures including anthropogenic forcing on the one hand, and the temperature calculated without anthropogenic forcing on the other hand. The anthropogenic temperature rise usually quoted as being of one degree over one century, is indeed of about one degree, but over about thirty years.

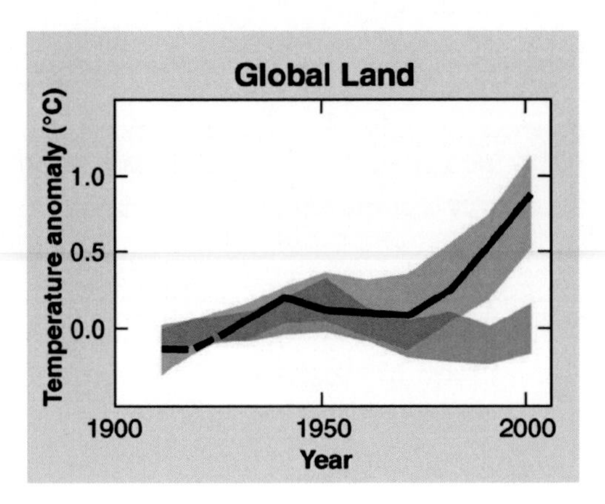

Fig. 5.12. Global land average temperature: measured (blackline), calculated with natural and anthropogenic forcing (pink band) and with natural forcing only (blue band). In this blow up of the central lower graph of Fig. 5.10(a), the departure of the pink and blue bands since the mid-seventies is clear. Up to that time, there is no clear evidence that anthropogenic forcing has contributed to a temperature rise.

The rapid change in the global temperature evolution since 1975 appears also on a more local scale. Figure 5.13 shows that evolution in Switzerland as recorded by Meteo Suisse. It shows recorded temperatures from 1880 up to 2007, taking the average temperature over the period 1961 to 1990 as a reference. There was a slow warming by about half a degree from 1880 up to 1980, followed by a much faster one of about 1 degree in the last 30 years.

Both global and local temperature variations suggest that since 1975 we may have entered into a regime of climate instability. Up to that time, the biosphere was remarkably resilient to anthropogenic perturbations, showing no clear effect of greenhouse gas emissions on the temperature. This may explain in part the intensity of the controversies surrounding the reality of anthropogenic effects. But since 1980 the evidence of an anthropogenic effect is overwhelming. The built-in stabilizing mechanisms of the biosphere, still poorly understood, are now evidently unable to preserve climate stability.

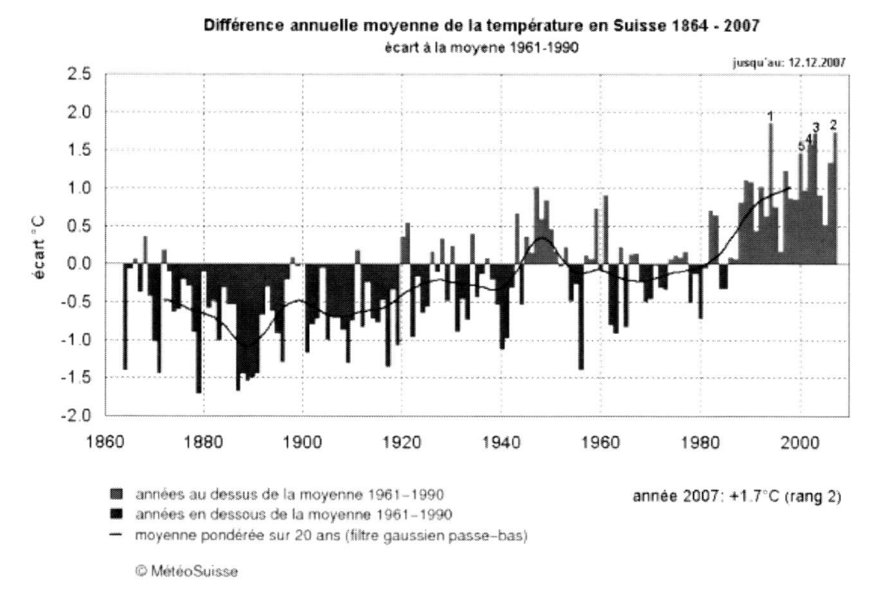

Fig. 5.13. Evolution of the temperature in Switzerland as reported by Meteo Suisse, from 1880 up to 2007, the reference being the average temperature over the period 1961–1990. Note the rapid increase since 1980, as compared to the slow rise during the period 1880–1980. The continuous line is calculated taking a 20-year average.

5.3.2.2. *Expected temperature rise in the 21st century*

If we take the year 1975 as the reference year to estimate the rate of temperature change, we obtain a figure of one degree over 30 years. If we extrapolate this rate of change, we get about 3 degrees over one century. Temperature at the end of the 21st century should be about 3 degrees higher than it is today. This is of course a crude estimate, but it is in line with the much more sophisticated predictions contained in the IPCC report.

Computer simulations quoted in this report for different development scenarios, indicate a further temperature increase during the 21st century ranging from half a degree to 2 to 4 degrees depending on the rate of greenhouse gas emissions. Margins of uncertainty for these simulations are such that temperature elevation values can vary from 1 to 6 degrees. One reason for the large margins of uncertainties is a feedback mechanism between temperature rise and greenhouse gases concentration. At higher temperatures CO_2 land and sea intake is reduced, more gases remain in the atmosphere, which leads to stronger temperature elevation and so on. The strength of this feedback effect varies among models leading to a rather broad range of predictions.

An important parameter is the time lag between variations in CO_2 concentrations and temperature evolution. For instance, Berger et Loutre have shown in their own simulations that by the end of the 21st century the temperature rise ranges from 1.5 to 3 degrees under assumptions for CO_2 concentration similar to those of IPCC, but they have also investigated how long it would take to return to equilibrium if at some point greenhouse gas emissions would stop. This might occur for instance if technology would at some point allow carbon dioxide storage, or if all available fossil fuels would have been burned out. They calculate that if emissions would stop in 200 years, it would take the system 800 years to return to equilibrium. There are even more severe possible outcomes. If for instance temperature elevation was so strong that it led to melting of the Greenland ice cap within 5,000 years, it would take the system 40,000 years to return to equilibrium.

5.3.2.3. *Consequences of further temperature rise: ice melting*

We have said that the last five interglacial periods, including the present one, are very similar in the sense that for all of them there is a perfect correlation between the temperature and the CO_2 cycles, with a temperature rise of 8 to10 degrees, and a CO_2 concentration rise of 100 ppm. Nevertheless, if we look at the fine details, some differences between the current and previous ones do appear. We have already mentioned one of them, which is that the temperature has remained stable for a longer period of time during the current period than it did in previous ones. Here we wish to point out to an additional difference which concerns the temperature increase. A close look at the Vostok data in Fig. 5.1 shows that in previous cycles it was close to 11 degrees and in the current one it has so far been only of 8 degrees. This difference of two to three degrees is in fact very important.

The melting of arctic ice and of glaciers in general has often been cited as a spectacular proof of global warming, the implication being that it is due to human activities. It certainly makes for impressing pictures. But geological records (quoted in the IPCC report) show that massive melting of arctic ice did occur in each of the previous interglacial periods. Certainly this was not due to human activities! Up to recent years, it is the current interglacial that has been the exception, with less ice melting than in previous interglacial periods. One can reasonably conjecture that this is due to the difference of 3 degrees between the maximum temperatures reached in the current and previous interglacial periods. The increase of temperature in modern times — whatever its origin — is only bringing the present interglacial in line with previous ones.

A consequence of the expected temperature rise of about three degrees during the 21st century will be the melting of Arctic ice, as it occurred in previous interglacial periods. It will lead to an increase of sea level of several meters, according to past records. Should the Greenland ice sheet melt, the sea rise would be of several 10 meters, but this is not expected to occur from available simulations.

So far we have discussed aspects of global warming over which there is a rather broad consensus. The rough estimates for the future evolution of the average temperature that one can make based on a careful examination of the data, as we have done, are not far off the mark as compared to much more sophisticated computer simulations. But in a sense, global average effects are not our major cause of concern. Effects that vary in space and time are really our primary interest, but this is a far more difficult question.

5.4. Climate changes in space and time: back to entropy

After all, for many of us an average temperature increase of a few degrees is not a major problem, particularly since it is predicted that it is at northern latitudes that it will be felt most strongly. Temperatures a bit warmer in the summer and somewhat milder in the winter might even be pleasant for people living in Moscow, Berlin, Paris or London to give a few examples. Thus predictions concerning global average temperature changes are not by themselves terribly worrisome.

What we really would like to have are reliable predictions concerning weather evolution at a given place — where we live. And what we are worried about is not so much the average temperature evolution at this place, but by how much temperature and other climatic parameters such as precipitation, winds and so on will fluctuate and how frequent these fluctuations will be. We know from experience that such predictions are far more difficult to make since, at the present time, detailed meteorological predictions can only be made for about one week.

In broad lines, the IPCC report does however make two kinds of predictions:

(1) There will be more frequently extreme weather conditions. This includes heat waves, heavy precipitations and an increase in tropical cyclone intensity.
(2) There will be more precipitation at higher latitudes and less in subtropical regions, where there will be a decrease in water resources and a higher risk of droughts.

While the second prediction concerns a change in specific regions (higher and lower latitudes), the first one is of a more general nature in the sense that the location of the more frequent extreme events is not specified.

In other words, one of the predictions being made is that the weather, being less stable, will be less predictable.

This trend has already been noticed, for instance very strong rains and floods occurring sometimes so suddenly that populations could not be given enough time to get prepared (or evacuated), even in regions usually known as having a temperate climate, such as France for instance.

This is the place to go back to entropy. *In fine*, most of the energy contained in the fossil fuels ends up as heat rejected at or near ambient temperature as a result of irreversible processes. As we have seen in the previous chapter, the meaning of this "lost" energy is that there is an increase of entropy in the biosphere. As a result of irreversible processes linked to our activities, energy has been dumped into the environment that can never be recovered in any useful way. It has not disappeared, but entropy has increased.

We postulate that the rapid increase of the concentration of CO_2 in the atmosphere and of temperature seen in the last decades is a result of the total increase in entropy due to all the irreversible processes that took place when fossil fuels were burned massively. This increase was finally too large to be compensated for by the only available external energy input, namely solar radiation. An increase in entropy means an increase in disorder, here climate disorder. It seems that since the 70s onward, climate stability was affected. A world with increased entropy is a less predictable one. The more frequent extreme weather conditions predicted by the IPCC report are, we believe, the unavoidable consequence of a massive entropy increase in the biosphere.

5.5. The entropic meaning of sustainable development

A world of sustainable development would be a world where the entropy generated by man's activities would be held in check by the input of energy that the earth receives from the sun, as was the case up to a few decades ago. In this ideal world, there would be no anomalous increase of CO_2 in the atmosphere and no climate disorder. That does not mean that there would not be climate changes — they have been and they will remain an integral part of the history of the earth, as we have seen. But these changes would be predictable and therefore could be managed. On the contrary, increased climate disorder due to increased entropy cannot be managed because it leads to events that cannot be predicted.

These very qualitative remarks are related to the more general problem of climate stability. Records of past temperatures and greenhouse gases concentrations demonstrate that the climate on earth is stable. For instance, in the last 5 million years, each time temperature has risen, some mechanism brought it back down, and vice-versa, as can be seen for instance in Fig. 5.1. Some authors have pointed out that physical parameters alone do not lead to such climate stability. For instance, a rising temperature increases exponentially water evaporation from the oceans. Because water molecules constitute a greenhouse gas, this increase leads to further temperature increase, which may result in a run away process ending up with complete evaporation of the oceans. An inverse process could lead to complete glaciation of the earth.

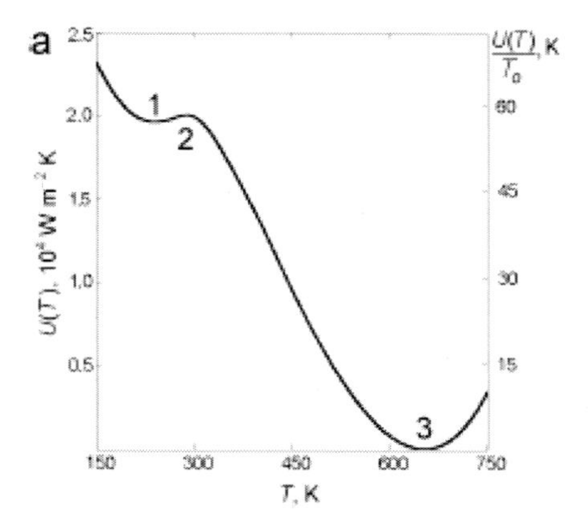

Fig. 5.14. Climates of complete glaciation (point 1) or complete ocean evaporation (point 3) correspond to stable minima of a potential function U. The current climate temperature (point 2) is an unstable state. (After Gorshkov et al., Atmos. Chem. Phys. Discuss. **2**, 289 (2002)).

According to some models (see Fig. 5.14), the states of complete evaporation (**3**) and complete glaciation (**1**) are stable states, while the current state where most of the water is in liquid state (**2**) is unstable. Biological regulation would in fact be the mechanism ensuring climate stability. These models, together with the current large entropy release, and *rapid* changes observed since 1975 in CO_2 concentration and temperature, may be cause for more serious concern than just a temperature rise by a few degrees.

5.6. Concluding remarks

We are currently in an interglacial period, which has already lasted longer than previous ones. Current climatic models cannot tell us how much longer it will last. It may end within a few thousand years, but it may also last longer, maybe up to 50,000 years. When it will end there will be a major climate change. If the past is any guide, glaciers will be back in northern Europe and northern America.

Climate change is a difficult subject. Because it is not completely understood, it is not easily summarized, and a fortiori reliable predictions are hard to make. Yet, we have come to some conclusions that may serve us as a guide for future entropy management strategies.

Anthropogenic effects have only become clearly noticeable since the 1970s. Since then, however, CO_2 atmospheric concentration and temperatures have risen at a rate that is unprecedented in previous interglacial periods. Temperature is rising at a rate of 3 degrees Celsius per century. CO_2 concentration is now 30% higher than it has been in any previous interglacial period. Further temperature rise by several degrees up to the end of the 21st century is unavoidable, leading to extended ice melting. Climate instability cannot be ruled out, as the rate of entropy release is now too large to be held in check by incoming solar radiation.

Taking the empirical evidence as a guide, one may conclude that in order to avoid climate instability greenhouse emissions should be limited to the level they reached in 1970, up to which time no anthropogenic temperature rise occurred. This means that they should be reduced down by a factor of 2 compared to current levels. This should be the minimum objective of an entropy management strategy, assuming that climate stability can be restored. If the climate has now entered a regime of strong instability, this objective may not even be sufficient.

Chapter 6

Fighting Entropy with Technology

On the billion year time scale, the biosphere is a world of increasing order as life spreads on the globe and assumes forms that are more and more sophisticated. This increasing order is made possible by a negative entropy balance, as solar radiation over-compensates entropy production due to unavoidable irreversible processes such as the decomposition of living organisms, atmospheric and oceanic currents and so on. A fraction of organic matter is even stored as coal, oil and gas. Finally a state is reached in the last few million years where the climate is rather stable if one includes regular periodic variations as recorded at the Vostok and other locations.

6.1. Motivation for fighting entropy increase: ensuring climate stability

This stable climate is as much the result of physical factors — solar radiation, interaction between atmosphere and oceans — as it is of biological processes. Without the spread of life, there would have been no climate. The close correlation between the temperature and CO_2 cycles in the Vostok records constitutes evidence of the strong link between physical and biological factors in determining climate changes. As we have seen, variations of physical factors only — for instance of solar insolation — are not sufficient to explain climate changes in the past. In current models, these changes can only be reproduced when CO_2 variations — the biological side — are introduced by hand. The exact way in which the physical and biological spheres interact and determine the climate is still to be worked out. Until this is clarified it will be difficult to understand how climate stability is achieved, and how this stability can be destroyed.

Comparison of past temperature records with the results of simulations that do or do not include anthropogenic effects, has revealed two new features that may be of importance concerning climate stability.

The first one is that up to the mid-seventies temperatures have remained remarkably insensitive to CO_2 anthropogenic emissions. The second one is that when anthropogenic warming started, it did so at an alarming rate. While the number usually quoted is of about one degree over one century, close examination of the data gives rather a figure of about one degree over the last 30 years.

Considered together, these two features — stability before 1975 and rapid temperature rise after that date — suggest that climate instability may have set in around 1975. We have proposed that this instability point was reached because entropy release, as measured through the rate of fossil fuel combustion, did pass a certain critical level. If this assumption is correct, it is insufficient to consider different scenarios that relate in some continuous way the amount of emitted CO_2 with temperature elevation. If we have passed a critical point, changes may be faster and stronger than continuous models predict.

To illustrate what is meant here, the stability of a complex system can be represented by a graph giving its potential energy as a function of a parameter P that can combine many factors. Figure 5.14 was an example of such a graph, where the parameter P was the temperature. A local minimum represents a stable state: a small fluctuation in the controlling parameter P will leave the system near its equilibrium point, where it will return to by itself once the fluctuation is over. But a strong fluctuation in P can take the system over the hump, from where it can go down to a completely different equilibrium point. Figure 6.1 is a possible extension of Fig. 5.12. Point **A** represents a regime of glaciation, point **C** a regime of very high temperatures, and point **B** is a local minimum where climate stability with a large mass of liquid water is ensured by the interaction between physical and biological factors: this is where we are now. The parameter P in this graph might be total rate of entropy release. Models valid near point **B** cannot predict what may happen once the system goes over the hump either towards point **A** or towards point **C**. Once over the hump, the system may not go back to point **B.**

The main reason why we should drastically cut on our use of fossil fuels is not only that we would like to conserve them for future use. Running out of fossil fuels should not be our main concern. Our major objective should be to reduce entropy release, with the hope that this reduction will prevent a climate run-away. In this chapter, we shall mention various means to reduce entropy release without affecting drastically our way of life. A more drastic approach — a world where fossil fuels would not be burned anymore — will be envisioned in the next chapter.

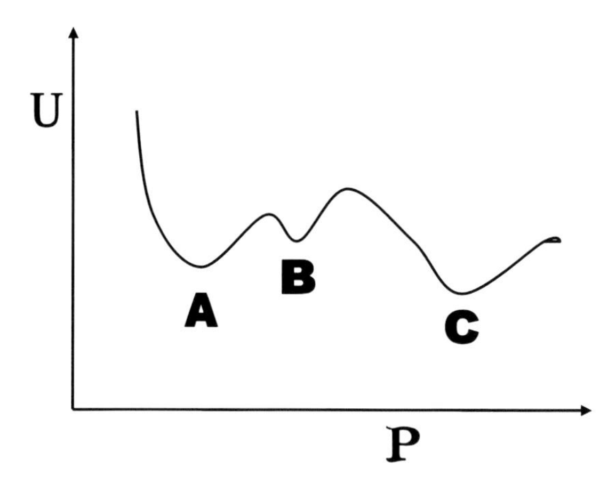

Fig. 6.1. Possible stable climate points. At point **A** there is complete glaciation, at point **C** complete evaporation of oceans. At point **B** oceans are in liquid state and stability is ensured by interaction between physical and biological factors.

6.2. By how much do we need to reduce anthropogenic entropy release

Since most of the entropy released has been so far through the combustion of fossil fuels, we can take CO_2 emissions as a measure of that release. At the present time CO_2 emissions from fossil fuel combustion (including coal, oil and gas) are equivalent to about 7 Gigatons of carbon per year (one Gigaton is equal to one thousand Million Ton), see Fig. 5.8. How can we decide whether this is too much or not?

Photosynthesis fixes about 100 Gigaton of CO_2 per year. In a stationary state CO_2 fixation is equal to CO_2 production by organic decomposition: 100 Gigaton per year is the biosphere natural entropy turnover. Current CO_2 emissions by fossil fuel burning are therefore about 7% of that turnover, which is not an insignificant amount. The rapid increase of CO_2 atmospheric level during the last three decades is an indication that the physical and biological mechanisms of the biosphere are not able to absorb this level of emissions.

Total energy consumption per capita in western societies is about 5 times larger than the average per capita consumption of the earth's entire population. This means that if the entire world's population would consume energy at the same level as people in western countries do (assuming that such energy is available), the total entropy release by mankind would become more than a third of the entropy turnover of the biosphere excluding mankind. One can doubt whether climate's stability, as shown schematically around point B of Fig. 6.1, would be maintained.

Up to 1975 carbon emissions from fossil fuel combustion were less than 4 Gigaton of Carbon per year. Since anthropogenic effects on the climate were not noticeable before that date, one may empirically conclude that this level of emission, and the corresponding level of power spent and entropy released, is acceptable. These 4 Gigaton correspond roughly to 1000 W of power spent per capita, or half of the global average power spent today. The other half would have to come from other, preferably renewable, energy sources.

6.3. Entropy management strategies

It should be clear that the objective of entropy management strategies should not just be to reduce the use of fossil fuels, but more generally to maintain entropy released by mankind at a level possibly comparable to what it was up to the mid-seventies. Strategies of entropy management can be divided into two categories. The first one consists in reducing as much as possible the degree of irreversibility in processes of energy conversion. The second one is to generate the energy that we use in such a way that it does not affect the total entropy balance. Let us give some examples in each category.

6.3.1. *Minimizing irreversibility processes by developing improved technology I: the motorcar*

When we drive a conventional motorcar, powered by an internal combustion engine, several irreversible processes result in entropy release. Some are due to factors that limit the conversion efficiency of the engine, which is of about 20%. Others are related to the use of brakes when slowing down the car, either to prevent it from accelerating dangerously when we go down a slope, or because traffic requires it. Heat is then produced and dissipated in the atmosphere. It is energy that can never be recovered, which means entropy released. Other irreversible processes resulting in entropy release include friction in the transmission and friction between the moving car and the surrounding air. There is also continuous deformation of the tires.

If there were no irreversible processes taking place when driving a car, we could go to work and come back home without releasing any entropy, which means that we would not consume any energy. For instance, energy spent and converted into potential energy when driving up a slope would be regenerated when driving down that same slope, instead of using brakes to slow down the car. Similarly, energy spent to accelerate (acquiring kinetic energy) could in principle be converted back into its original form when slowing down.

But regeneration cannot be performed if our primary energy is provided by fuel combustion. No on-board practical machinery can give us back the burned fuel, because it was generated under very special conditions of temperature and pressure several hundred million years ago.

But regeneration is possible if we use other forms of energy as our primary on-board source. The reader has probably realized at this stage that some cars do already, to some extent, reduce irreversibility processes precisely by slowing down the car with an electrical generator that transforms the kinetic energy of the car into electricity that is then stored in a battery, and can be used again later on (this is what is done in hybrid cars where the electrical motor is also used as a generator when slowing down).

The internal combustion engine is the main problem of the conventional motorcar because it does not allow regeneration of the energy spent for acceleration or going up slopes.

6.3.1.1. *The all-electric motorcar*

One obvious solution is the all-electric car. Not only because it is cleaner (zero emissions by the car) and "saves" on oil consumption, but more fundamentally because it allows a reduction of irreversible processes by *re-generating* electricity, storing it and using it again. Hybrids are an important improvement in the right direction but are only a partial solution.

Since all of this has been known to engineers for many years, one may wonder why the all-electric motorcar is not manufactured on a large scale. Worst, why did all attempts to put on the road electric cars end up to now in commercial failure, so that we are still driving cars powered by internal combustion engines?

Up to the beginning of the 20th century there was still some competition between internal combustion engine powered cars and electric cars, but the latter quickly disappeared. There were two reasons to this defeat of the electric car. First of all, the heavy weight and short life time of batteries. Second of all, the fact that the price of gasoline went down dramatically as oil industry developed. This took away any incentive to solve the battery problem. Besides, pollution due to the use of internal combustion engine powered cars has only recently become an issue. Increasing gasoline price and the fight against pollution may now give the all-electric car a second chance.

6.3.1.2. *The battery problem*

It may be interesting to have a closer look at what the battery problem exactly is. Let us see how much electrical storage we need to power a car. Typically with one liter of gas we can drive 10 kms in a mid-size car, and may be up to 20 kms in a small car. Energy stored in one liter of gasoline is in rough numbers equivalent to a power of 10 kW spent during one hour, or 10 kWh. Due to the low conversion efficiency of the internal combustion engine, out of these 10 kWh only 2 kWh are actually transformed into mechanical work. So we need, for a small car, about 1 kWh of stored electricity to drive up to 10 kms, assuming that the electrical motor has an efficiency close to 100% (a bit optimistic, but its efficiency is actually quite high).

These are easy numbers to remember: 1 kWh for 10 kms. In order to drive a practical distance of say 100 kms, we would need to store 10 kWh. Now the electricity stored in a typical lead-acid battery of the kind used in current motorcars to get the engine started and other functions is of about 0.5 kWh. We would therefore need 20 of those batteries to drive our 100 kms. But the electric car offers the possibility to re-generate electricity, store it and use it again. This may give us up to a factor of 2. So we are down to maybe 10 standard batteries. This sounds workable.

But here comes the serious battery problem: its life time goes down drastically if we "empty" it to the last "drop" of electricity before re-charging it. Repeated deep discharge is a battery killer. If we want batteries to have a life time comparable to that of the car itself, we may only allow a discharge of 10% of the nominal battery capacity during our 100 kms drive. Here we lose a factor of 10 — we need 100 batteries (and their 1000 kilogram of weight). This does not sound workable.

Fortunately serious work on developing lighter and more robust batteries has now started on an industrial scale. Current hybrid models can already be driven on electricity for a few kilometers. The next generation, using improved batteries, may allow all-electric driving for a few 10 kms. Batteries have now become a hot item. There is good reason to be optimistic about further progress. Commercially successful all-electric cars for urban and sub-urban driving, where the ability to re-generate electricity in heavy traffic is of greatest advantage, may soon become a reality. From an economic standpoint they are already quite attractive. As we shall discuss in the next chapter, running a car on electricity is now a lot cheaper than running it on petrol.

6.3.1.3. *The hydrogen car*

Much has been written on the hydrogen economy. The hydrogen car is one of its popular features. It uses fuel cells to perform a reversed electrolysis where hydrogen is combined with oxygen, electricity is produced and water vapor is released. The fuel cell has a higher conversion efficiency than the internal combustion engine. It has also some of the advantages of the all-electric car such as CO_2 zero-emissions (although water vapor is also a greenhouse gas). But on the other hand regenerating hydrogen by electrolysis of the released water is more cumbersome that regenerating electricity in the all-electric car.

One could of course consider a hydrogen-electrical hybrid.

It is interesting to note that the first patented internal combustion engine, invented by a Swiss citizen by the name of Francois-Isaac de Rivaz (French patent 1807), did burn hydrogen. A reproduction of his invention is shown Fig. 6.2. Hydrogen is stored in a reservoir seen at the rear of the vehicle, presumably under pressure. The engine consists of a piston pushed upwards in a cylinder by the combustion of hydrogen. Depending on the state of progress made on batteries and fuel cells, hydrogen storage may or may not be preferable to electricity storage in the long run.

Fig. 6.2. The hydrogen car of Francois-Isaac de Rivaz, French patent 1807. The hydrogen reservoir is to the rear of the "car" (Fondation Pierre Gianadda, Martigny, Switzerland).

6.3.1.4. *But where will the electricity for the all-electric car come from?*

Let us indeed assume that the automobile industry will be successful in developing the new batteries necessary for the all-electric car. There will then be an increased demand for electricity. Where will it come from? And in the first place did we gain anything by shifting entropy release from the internal combustion engine to the power plant that generates electricity?

This question has been hotly debated, but in the end the answer is rather clear. If one compares the total losses that occur along the way, from the oil well to the moving car, it turns out that they are twice as high for the conventional car as they are for the electric car. This difference comes about mainly because the conversion efficiency of a fossil fuel central power plant is more than twice that of the internal conversion engine. It is therefore more efficient to convert first the fossil fuel to electricity, transport it, and store it in on-board batteries, than to produce gasoline and burn it in the internal combustion engine. In short, we shall release less entropy, less CO_2, and less pollution with the electric car than with the conventional one.

This is why the increased use of electricity due to a massive use of the all-electric motorcar may in the end not be that large. The strong reduction in irreversible processes made possible by the all-electric car reduces considerably its net energy consumption. In addition, the all-electric car may use electricity produced from primary sources different from fossil fuels, such as renewables or nuclear power plants. Many more options will be available, which will in itself be a positive development.

6.3.2. *Minimizing entropy production by improving technology II: space heating and cooling*

As a second important example, let us see how homes and buildings can be maintained at a comfortable temperature with a reduced entropy release. This is quite an important item since about a third of our energy consumption goes into doing just that (a third also goes into transportation, which is why we have discussed at some length the example of the motorcar).

The traditional way to keep a comfortable temperature in a building is to burn a fuel (coal, oil, gas, wood) to heat up a fluid (water or air) and to circulate it through the building. Energy spent in the process cannot be recovered — the end result is that entropy is dumped into the environment. Alternatively, electricity can be used to power a heat pump with the same end result in terms of entropy production. Cooling is in general done by air-conditioning, which is also a heat pump but working in the opposite direction as heat is pumped out of the building and dumped outside.

Various schemes have been considered to eliminate or reduce energy use for heating and cooling.

6.3.2.1. *Two types of solutions: improved insulation or increased thermal mass*

Improving insulation is at the moment the accepted way to reduce the energy required to keep a building at a constant temperature. It is clear that if our home could be perfectly insulated thermally from the surroundings it could be kept at a constant temperature at no cost in energy, except of course that one would need to compensate for the amount of energy dumped as heat inside the house because of one's presence and the use of various appliances. Otherwise one would not be affected by outside low temperatures in the winter and high temperatures in the summer. No need to heat in the winter (you would only have to turn on a heat leak, such as opening an orifice to the outside world for a few moments from time to time), and no need to cool in the summer.

Of course there is no perfect thermal insulation, so we need to define insulation quantitatively. Let us define the temperature outside the house as T_{out} and the temperature inside as T_{in}. We define the thermal conductance K through the walls and roof of the house as the ratio of the number of calories that flow in (or out) per unit time (let us say seconds), to ($T_{\mathrm{out}} - T_{\mathrm{in}}$). If the thermal conductance is zero, $K = 0$, we have a perfect insulation, no calories will flow in or out and T_{in} will stay constant whatever T_{out} is.

Now there is a different way to keep temperature constant, which is to live in a house that has a very high heat capacity, which for illustrative purposes we call thermal mass. The heat capacity of a body C is the ratio of the amount of heat Q transferred to it divided by the resulting temperature increase ΔT. If the heat capacity of the body is infinite, its temperature will not change irrespective of how much heat it receives (or loses). In a sense, this is what our ancestors did when they lived deep inside caves several ten thousand years ago, before our nice current climate set in and made possible other solutions.

6.3.2.2. *Keeping the temperature constant is a question of time scale*

The reader might have recognized that the ratio of heat capacity to thermal conductance has the dimension of a time. This is because both quantities involve a ratio of heat to temperature, but in the case of thermal conductance it is more exactly a heat flow, namely heat per unit *time*. So let us call this time, the ratio of the heat capacity of the building to the thermal conductance through its walls, roof, door and windows, the thermal relaxation time and denote it by the letter τ. If the heat capacity of the house is infinite, or if thermal conductance through the walls is zero, the time τ is infinite. This means that any temperature difference can be maintained for ever between the outside and the inside.

Fortunately, this is really more than what we need. All we need is that this time should be long enough. A minimum requirement is that it should be longer than the day-night cycle, so that the building will not cool down significantly at night during the winter and will not warm up too much during the day in the summer. Ideally, this time should be of several months so as to smooth out the winter-summer contrast.

Houses of rich people (kings and the like) that were built centuries ago had very thick walls and small openings, as we can see when we visit their old castles. This combination of a large heat capacity and small thermal conductance gave them the required long relaxation time. Although this way of living had been greatly improved compared to that of our ancestors of older times who used to live in caves, it used the same combination to achieve a long relaxation time.

In fact, we do not need to visit old castles to experience living spaces where temperature is kept constant in a purely passive way. A good wine cellar does precisely that. Its temperature is about the average between winter and summer, which turns out to be in many places in the range of 10 to 15 degrees Celsius — exactly what wine needs.

6.3.2.3. *Towards zero entropy release buildings*

Living in a wine cellar is not very attractive anyhow, but in addition a temperature of 10 to 15 degrees is not comfortable either. So we need to do better, without losing sight of what our main objective is i.e. trying hard to maintain a comfortable temperature by purely passive means. A good way to start to evaluate the thermal properties of your house is to perform an easy and cheap experiment, which I recommend the reader should do (experiments are always fun, but in addition this one is quite useful).

Select a reasonably cold evening in the winter, measure the outside and inside temperatures and then, before you go to bed, shut down your heating system. When you get up in the morning, measure again the inside and outside temperatures. The difference will now be smaller than what it was in the evening: the house has cooled down. If the difference has not changed much, your house is well insulated. If it has changed a lot, you should do something about your insulation.

To be specific, suppose that the outside-inside temperature difference was 20 degrees in the evening, and it is 18 degrees or more in the morning. Your house is very well insulated. If it is between 15 and 18 degrees, it could be improved but is still OK. If it is 10 degrees or less, the insulation of your house is a disaster. You could save a large fraction of your heating bill by improving it. No need to call on an expert to evaluate how well insulated your house is, this is really a do it yourself and guaranteed evaluation.

Of course, you will need the help of an expert if the test has turned negative for your house. It will be his task to make the most appropriate recommendations for improvements. After they will be completed, I recommend that you make the same measurement again. This will allow you to make a precise evaluation of the expert's work.

Nowadays the emphasis is on reducing the thermal conductance by using new insulating materials, rather than increasing the thermal mass. However it is all a question of what the climate is where you live. If you live in the Middle East you are lucky: the average temperature between winter and summer is close to the comfort norm of 20 degrees. You can choose any combination you wish between a large thermal mass and a small thermal conductance to obtain a large relaxation time. Living in a cave would be perfect, except of course for the absence of light. But if you live in Northern Europe or Northern America where the average year round temperature is less than 15 degrees, the emphasis should be on a small thermal conductance because you will need some heating and if you choose a large thermal mass it would have to be heated as well, which would be quite expensive.

The quality of thermal insulation is expressed in watt per square meter per degree of difference between the outside and the inside temperature. With modern insulation it can be as low as 1 watt per square meter per degree of temperature difference across windows, and 0.2 across walls. This is four times smaller than in houses built in the 60's, and the amount of energy spent on heating needed to go through the winter is accordingly 4 times smaller. It can in fact be so small that very little heating is needed, if at all. For instance, with large windows facing south the amount of solar radiation received during the day may provide most of the heating needed. Except in the most difficult climates it is becoming possible, with passive house climate control only, to keep a comfortable temperature all year round. For instance, thick rotating concrete walls having a black painted side and a white one can, in the winter, absorb heat during the day (black side facing the outside) and release it by radiating it back to the house during the night (black side facing the inside). Vice versa, in the summer white will face the outside during the day and the inside during the night, releasing towards the cool sky heat accumulated during the day. Air circulation is also controlled so that in the winter cold air flowing in is pre-heated by warm air flowing out. In the summer colder air circulation during the night can effectively remove heat accumulated during the day.

In some regions such as the Mediterranean basin, where there are no extreme weather conditions either way, passive heating and cooling techniques can provide most of the temperature control. In regions where there are more severe weather conditions, improved thermal insulation can reduce the needed amount of energy by a substantial factor, often quoted as being of the order of 4, such as offered by the Minergy standard in Switzerland (the Minergy standard is that a thousand liters of fuel are sufficient to heat a 170 square meter built area detached house through the winter). This represents a major improvement towards the ultimate goal of zero entropy production buildings.

6.3.3. *Reducing entropy release in industry*

We have given two examples of entropy release reduction, one in the domain of transportation and the other one in residential heating and cooling. The third major area of energy consumption is industry, particularly heavy industry. Energy is used for instance to heat materials up to elevated temperatures (such as for metal production), or to produce heavy mechanical work. Heavy industry is possibly the area where reducing energy consumption is most difficult. It is also a very technical domain, well beyond the simple approach that we have been using here, but advanced technology such as improved furnaces and computer controlled processes can reduce losses. Where heavy mechanical work is done, electrical motors with low-loss superconducting windings can be useful. In general, the use of superconductors in the power sector can be very beneficial in the long term.

6.4. Energy generation impact on global entropy release

The other side of an entropy management strategy is to develop energy generation and distribution techniques that tilt the total entropy balance as little as possible.

6.4.1. *Energy generation from renewable sources*

Renewables, as they are now commonly called, are in general defined as sources of energy that will remain available for ever, contrary to fossil fuels for instance. In addition, they are supposed not to have a negative impact on the environment. Said in the language that we have been using here, these are sources of energy whose exploitation should not increase entropy release.

6.4.1.1. *Biomass*

Cutting wood and using it for heating and cooking is the way our ancestors used biomass. Some of us still do so.

In a steady state, trees grow and eventually die, then decompose into their original constituents. CO_2 that was taken from the atmosphere is restituted to it, and a new cycle goes on. The entropy balance is zero. Now suppose that when a tree dies you burn it instead of letting it decompose. From the entropy cycle point of view it makes no difference with the previous case where you let the dead tree decompose. The same amount of CO_2 will be released into the atmosphere, and will be removed from it later by photosynthesis. By extension, you can even cut trees and burn them before they die on the condition that you take care to plant or let new trees develop at the same rhythm so as to leave the total forest mass unchanged. As long as the total entropy is unchanged by your actions, the environment has not been modified. Of course if you destroy the forest either in part or in totality, you will have modified the original cycle in an irreversible way and increased entropy because less CO_2 will now be removed from the atmosphere.

Ideally, biomass obtained from wood could be used for our energy consumption without generating entropy. Note however that even in this best of all possible words, some entropy will in fact be produced because you will need entropy generating mechanical tools to cut trees (you must have noticed how a saw heats up when you cut a piece of wood), and later to transport them and condition them. But properly managed this is likely to be a small effect.

Biomass grown to produce biofuels is altogether a different story: it is now well understood that it can have devastating effects on the biosphere, because it leads directly or indirectly to deforestation. What is going on in different parts of the world is that forests are being destroyed to make room for the growth of plants for biofuels production on a massive scale, or to grow plants such as soybeans that were previously grown on existing arable land where biomass (corn for instance) is now cultivated for biofuels production. In addition to the resulting increase in entropy, the competition between biofuels and food production is also driving up food prices in an uncontrolled way.

6.4.1.2. *Solar heating*

Solar radiation that hits the surface of the earth is partly reflected and partly absorbed. Leaving aside the part absorbed by photosynthesis, the light absorbed by water, rocks, sand and so on, just warms them up a bit: the energy brought by sun rays is now lost in the sense that one cannot use it anymore.

Now suppose that you want to build a device that absorbs light very efficiently and can transfer the collected heat to where you need it. For example, you take a metal plate, paint it in black, and solder to it water pipes. To complete the device, you enclose it inside a box with a glass window facing the sun. This glass window has two important functions: it prevents hot air in contact with the black metal from escaping into open space, and it reflects back the infrared radiation emitted by it (greenhouse effect). Water can now be circulated through the warm pipes (typically at 60 degrees centigrade), stored, and when needed used for showers or other uses. You have just built a hot water solar collector and storage. The device can be made more effective by using a special black paint, called selective black, which absorbs very effectively solar radiation in the high frequency part of the spectrum (red to blue) but emits very little in the infra-red part. To avoid freezing problems in the winter, water can be replaced by another fluid such as glycol that does not freeze and will be circulated through a heat exchanger in a close circuit to heat up the water you need.

From our entropy standpoint, this hot water solar system is almost ideal: it makes no difference whether solar radiation would just have hit our house roof and be transformed there into useless low grade heat, or whether it first heats up water to a higher temperature before being eventually degraded. The global entropy balance has not been modified, just as in the case of biomass. But again here, nothing is ideal. To fabricate our device we had to use a metal plate, water pipes and the glass window all manufactured using relatively high temperatures processes, which means entropy released. One should calculate how much entropy was released in this fabrication process and compare it to the entropy that would have been releaseded if we had heated our domestic hot water by burning fuel or by using an electrical heater. It turns out, like in the case of biomass, that the balance is very favorable.

Hot water solar collectors are very practical devices. They have been in massive use in some countries such as Israel (where selective coatings were first applied by Tabor) and Japan for many years, and their use is now spreading to other countries.

6.4.1.3. *Thermal solar electricity*

The same principle is used in high temperature solar collectors for the production of steam and electricity. To reach higher temperatures, solar radiation is concentrated, for instance through parabolic troughs whose axis is oriented north-south. Their inclination tracks the sun during the day. Sun rays are focused unto a metal pipe unto which a highly selective coating has been deposited. This pipe is enclosed inside an evacuated glass tube, so that heat cannot be transferred directly to the glass tube via hot air. A fluid, for instance oil, is circulated through the metal pipe and brings the heat to a heat exchanger inside a boiler where steam is produced. Electricity is generated through a turbine and a generator, just like in a conventional power plant.

Again, from our entropy standpoint, this is an excellent device. It makes no difference whether solar radiation is directly transformed into low grade heat in the field, or whether it is first converted into electricity before being downgraded somewhere else. But again, the device is not ideal since high temperature processes had to be used to produce the mirrors, the pipes and their coating, and the glass tubes. But, compared to our domestic hot water solar collector, the thermal solar electricity plant has an important new property: it can self-reproduce. A fraction of the electricity generated can be used for the fabrication of parts that require high temperature processes. This self reproduction capability offers some similarity with that of biomass, but a solar thermal electricity plant has over a forest immense advantages. It has higher conversion efficiency (about 30%), needs no water and can be installed in deserts.

It is interesting to note that this kind of power plant is not new. The first solar powered steam engine was invented by the French mathematician and physicist Augustin Bernard Mouchot about 150 years ago. Mouchot was a high school teacher in Tours who developed a number of solar devices, including solar cookers that were to be used in the French colonies at a time when there was a shortage of coal in France because of deteriorating relations with England. Mouchot wrote: "Eventually industry will no longer find in Europe the resources to satisfy its prodigious expansion...Coal will undoubtedly be used up. What will industry do then?". He even conceived of the possibility to use solar power to produce hydrogen. Later, a power plant using the same mirror geometry and a black pipe inside a glass tube as described above, producing 73 kW, was installed in Egypt in Maadi near Cairo by the *Sun Power Corporation* at the beginning of the 20th century (Figs. 6.3a and 6.3b).

Principles of modern solar thermal power plants have remained the same as those of these earlier power plants, but the conversion efficiency has been greatly increased thanks to a number of new high-tech elements, which have turned the idea into a practical proposition. They include the use of advanced selective coatings

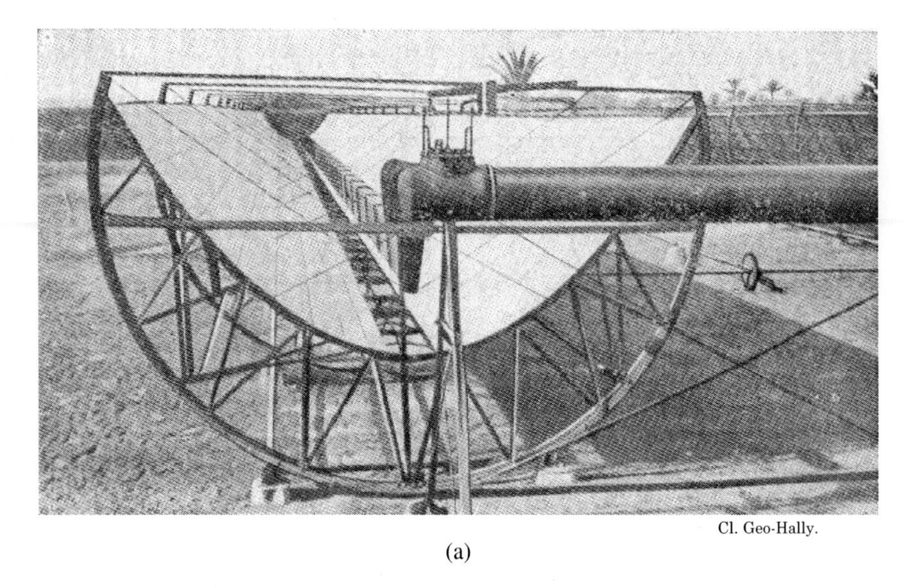

Cl. Geo-Hally.

(a)

Cl. Geo-Hally.

(b)

Fig. 6.3. (a) The parabolic trough installed in Maadi (Egypt) at the turn of the 20th century. Sun rays are concentrated on the pipe located at the focus of the parabola (in Le Ciel, Eds Larousse, Paris 1923). (b) The solar heated steam power plant at Maadi (in Le Ciel, Eds Larousse, Paris 1923).

that absorb 96% of the incoming radiation and re-emit in the infrared only 10% of the absorbed energy, as well as long lasting sealed vacuum tubes that have substantially reduced replacement and maintenance costs. As a result of these technological improvements the fluid temperature can be raised up to 400 degrees allowing the conversion efficiency to reach 30%.

Fig. 6.4. The Luz–Solel solar thermal electricity field in the Mojave desert, California. The scale of the field, now generating more than 300 MW, gives an idea of the progress accomplished since the Maadi plant was installed (Courtesy of Solel, Beit Shemesh, Israel).

Because they need concentrated sunlight, thermal solar power plants can only be effectively used in regions where skies are very clear most of the time. Diffuse light is inappropriate for them. In addition, single axis orientation as currently used in most solar fields limits locations to latitudes lower than 40 degrees, otherwise too much radiation is hitting the parabolic collector at an unfavorable angle. In practice, since the equatorial band between –20 and +20 degrees of latitude is very cloudy, the useful latitude bands lie between (–40, –20) and (+20, +40). In the Northern hemisphere, Southern-California, Mexico, the Mediterranean Basin, Middle East countries, Iran, parts of China, and in the Southern hemisphere Chili, South Africa and Australia are favorably located. The energy pay-back time of modern solar thermal electricity power plants — the time it takes the plant to generate enough energy to build another solar power plant — is of about 5 months. The pay-back time is an important and useful figure of merit since it is independent of the financial aspects of setting up the plant such as interest rate and tax incentives.

There exist other more strongly concentrating schemes using two-axis parabolic mirrors, or fields of mirrors concentrating sunlight on a receiver mounted on a tower, that allow higher temperatures to be reached and accordingly higher conversion efficiencies.

6.4.1.4. *Photovoltaics*

While thermal solar power plants are based on principles that were already known a very long time ago (remember Archimedes and his concentrating mirrors), photovoltaic solar cells are the outcome of modern science that was totally unknown one century ago and has only become fully understood half a century ago. The invention of photovoltaic solar cells is intimately related to that of the transistor, both being based on the properties of semiconductors.

As their name indicates, semiconductors are intermediate materials between metals such as Copper that conduct well electricity because electrons move freely, and insulators where electrons are strongly localized. Light of frequency ν incoming on a semiconductor can set electrons free if the photon energy $h\nu$ (h is the Planck constant, see Chapter 3) is sufficiently large to allow localized electrons to jump up to the energy of a state where they are mobile. For a semiconductor such as Silicium, the required frequency corresponds to that of visible light. Once set free, an electron leaves an empty state, called a hole, which can also move.

The electron has a negative charge and the hole a positive one. They will accordingly move in opposite directions when submitted to an electric field. A photovoltaic cell is a device with an internal electric field built-in: it accelerates electrons and holes in opposite directions towards opposite electrodes where they are collected. A difference of potential then sets in between these electrodes and the device becomes a "solar battery" that can provide current to an external load. The conversion efficiency of commercial photovoltaic panels based on crystalline Silicium cells is approaching 20%.

The beauty of photovoltaic solar cells (PVs) is that they provide high grade energy (electricity) without light concentration. As such, they can be operated in regions where skies are not so clear and a substantial fraction of the incoming radiation is in the form of diffuse light. Naturally, the amount of electricity generated varies in proportion with the total amount of incoming radiation, which will be as a rule smaller at northern latitudes. But this is not an "on-off" situation as for concentrating devices.

PVs were first developed for space applications, and remain the main source of power in satellites and space stations. There, weight considerations rather than cost are of primary importance. For this reason, Silicium has progressively been replaced for space applications by more advanced semiconductors such as Gallium Arsenide that give a higher conversion efficiency of about 30%.

From our standpoint — entropy — PVs suffer from the fact that their manufacture requires high temperature treatments to melt, purify and crystallize the semiconductor (mostly Silicium today). In spite of improvements in conversion efficiency and manufacturing processes, the pay-back time of Silicium PV cells is longer than that of thermal solar electricity plants. It varies from 2 to 4 years depending on location. Yet this longer return time is much shorter than the PV life time (over 20 years).

Different semiconductors and manufacturing processes are being actively studied to reduce the pay-back time, such as reducing the amount of semiconductor necessary for photovoltaic conversion. For instance, light is absorbed far more efficiently in

Gallium Arsenide than in Silicium. In principle, a thickness of a couple of microns of this semiconductor would absorb light as well as 100 microns of Silicium. Savings in the use of material translates in a reduced entropy release during device fabrication. Amorphous thin film PVs are also being studied, as well as cells using organic materials that do not require high temperature processing. Light concentration is evidently also a possibility to improve the pay-back time. One may hope that more research will reduce it to less than one year in favorable climates, and to 2 years in less favorable ones.

6.4.1.5. *Wind turbines*

Wind turbines are the successors of windmills, whose history goes back to more than two thousand years as they were reported to be in operation in Persia for grinding grain in 200 B.C. Wind mills were extensively used in the Dutch country around 1200 A.D. The first windmill for electricity production was built in the USA by Charles Brush in 1888. Wind driven electricity generators producing around 10 kW of power were commonly used in US rural areas in the first decades of the 20th century.

Modern windmills usually called wind turbines come in units that produce several MW of power. Rotors are mounted on top of high towers that can reach up to a couple of hundred meters in height.

A wind turbine is composed of a rotor, a gearbox to ramp up the speed of rotation to produce alternative current at commercial frequencies, a generator and control electronics. Modern aerodynamics, higher performance materials, engineering and electronics have all contributed to transform the historical windmill into a high-tech power plant. For instance, orientation of the turbine axis and the angle of attack of the blades are computer controlled. This, together with the use of high resistance materials, has allowed the exploitation of high wind speeds. One of the reasons for building higher wind turbine towers is that wind speed can increase by up to 20% for each 10 meters of elevation in places where there is strong wind shear (because of friction, wind speed decreases near the earth's surface).

A wind turbine has no direct impact on the global entropy balance. Instead of being transformed into low grade heat by friction against the landscape, natural or man-made obstacles, wind's kinetic energy is transformed into electricity that will be degraded somewhere else. The denomination "wind farm" is thus appropriate. Of course, some entropy has been released during manufacture of the tower, turbine and accessories. But this entropy footprint is small, when measured by the pay-back time which is on the order of a few months.

On the other hand, a wind farm competes for land space with other uses. Wind turbines are not incompatible with agricultural activities, but may limit them. For instance, their presence will put constraints on aircraft crop-dusting. Wind turbines are not compatible with forests. They are noisy, so no one wants to live near them. These limitations will indirectly affect the entropy balance.

A relevant figure is the power that can be generated by unit land area. It is of the order of 100 MW per square kilometer. This is of the same order as power generated by solar thermal electricity power plants.

The main problem with wind power is its intermittency. Wind power is highly sensitive to wind conditions, as it varies as the cube of wind speed: as a result of wind speed variability, the average amount of power produced is several times smaller than the turbine rating. Additionally, wind speeds are not known in advance with great accuracy. Intermittency has an impact on the entropy balance. It requires that additional power plants be built in order to supply electricity at times where wind power is unavailable. Alternatively, locations that are favorable for higher and more constant wind power may be remote from those where power is needed. Losses and the corresponding entropy release will then occur as electricity is being transported from where it is produced to where it is needed. These issues are discussed at length in the next chapter.

6.4.2. *Non renewables: fossil fuels versus nuclear*

6.4.2.1. *Improving the use of fossil fuels*

Fossil fuels (coal, gas and oil), still provide most of our energy needs and cover all sectors: domestic and industrial heating, transportation and electricity generation. In the short and medium term, improving the use of fossil fuels through technology is therefore the most efficient way to reduce the damage that they cause to the environment.

Oil fired power plants are still in use, but new ones are not planned for the future. New natural gas fired power plants are still being built because gas is favored from an environmental point of view, but its availability and price in the long term is an open question. Therefore efforts made for improving the performance of fossil fuel fired power plants are concentrating on coal fired plants, the availability of coal being ensured for a few more centuries.

A new coal fired power plant is inaugurated every week in China. Even in eco-conscious Europe more coal fired power plants are being built. This movement cannot be stopped but the efficiency of new plants can be improved. The world average conversion efficiency of existing plants is equal to 32%. In conventional power plants, steam is heated to a temperature of 540°C. In modern plants, called "supercritical", where steam is heated up to a higher temperature (700°C), the conversion efficiency has already reached 45% and it is hoped that 50% can be achieved. For a given electrical energy output, this translates into a large improvement in terms of reduced coal consumption. Once all old coal fired power plants will have been replaced, the amount of coal saved may be equivalent to the reduction achievable with renewables.

The amount of entropy released by power plants can be calculated by comparing the theoretical Carnot efficiency (corresponding to zero entropy release) to the effective one. As we have seen in Chapter 4, the Carnot efficiency is equal to the difference between the boiler and condenser temperatures, divided by the boiler temperature expressed in degrees Kelvin (counted from the absolute zero of temperature, equal to −273.16°C). For instance, when steam is heated at 540°C, the Carnot efficiency is about 60%. An actual efficiency of about 30% is only half the Carnot value: 50% of the energy that could in principle be retrieved is in fact lost in the old plants. In a modern "supercritical" power plant producing stream at 700°C and having a conversion efficiency of 50%, much less entropy is released. The Carnot efficiency is then about 65%: only less than 30% of the energy that can in principle be retrieved is lost.

6.4.2.2. *Pros and cons of carbon storage*

Fossil fuels are considered the villain of climate change, because of the massive emission of greenhouse gases that results from their combustion. It has therefore been concluded that CO_2 storage is necessary to achieve the desired reduction in emissions.

On the other hand, reducing the use of fossil fuels by improving conversion efficiency is the only way to reduce the amount of entropy released for a given electricity production. From this standpoint, CO_2 separation and storage is counter-productive because energy must be spent for separation and storage. The overall conversion efficiency of the plant will be lower. Indeed, less greenhouse gases will be emitted but more fuel will have to be burned to produce the same power, and more entropy will be released. It has been estimated that the implementation of CO_2 separation and storage reduces a plant conversion efficiency by 30%. This means that 30% more fuel must be burned to obtain the same amount of electrical energy.

Modern power plants also incorporate devices aimed at reducing the emission of particles and unhealthy gases such as nitrogen and sulfur oxides. Unfortunately, this cleaning up act as well has an entropy cost, because it lowers the conversion efficiency of the plant as more fuel must be burned to operate the particle separation device while reaching the same level of power generation. It is said that in China, where generating the requested amount of electricity is the primary concern, installed cleaning devices are sometimes just turned off.

6.4.2.3. *Nuclear power plants*

Nuclear energy has been at the center of a seemingly endless controversy. On the one hand, its proponents claim that it is clean — there is no air pollution, no need to transport massive quantities of fuel from where it is produced to where it is burned, there are no safety concerns comparable to those affecting people working in deep and dangerous coalmines. On the other hand, its opponents claim that radiation problems remain unsolved, particularly when it comes to the disposal of radioactive spent fuel.

By and large, opponents of nuclear energy have had the upper hand in the last decades. Under their pressure, political bodies have hesitated between two courses of action: closing existing nuclear power plants at pre-determined dates, as decided in Germany, or more mildly, not building any new ones at all, as was done in France for a long period of time. A revival of interest for nuclear energy has occurred in recent years, possibly due to a combination of two factors: an increasing awareness that fossil fuel combustion poses great dangers to the climate, and the necessity to increase electricity generation one way or the other. Neither solar energy nor wind turbines constitute the universal answer to this problem.

Uranium is the fuel used in nuclear power plants. The mineral contains two isotopes: U-235 and U-238. Only the first is fissile, which means that it disintegrates spontaneously and emits a neutron. At each disintegration, energy is released. If captured by another U-235 nucleus, this neutron will cause its disintegration. A chain reaction is possible, releasing much more power than in spontaneous disintegration; it will take place if the fraction of successful capture events is sufficiently high. This can be achieved either by slowing down the neutrons, or using a U-235 enriched fuel: in both cases the capture probability is increased. In both cases, the fraction of U-235 decreases with time and after a while the chain reaction will stop (or become too slow to be useful).

Although it still contains a substantial fraction of U-235, the fuel has been spent and has become a "radio-active waste".

Nuclear power plants built in the 80's are of that kind. It is estimated that they burn effectively only on the order of one percent of the precious U-235 isotope. If built on a massive scale such plants would quickly exhaust Uranium ore resources (the time scale is about the same as that for oil or gas). In addition, they would generate an enormous amount of waste, whose radio-activity lasts for tens of thousands of years. From the standpoint of their efficiency and waste (entropy again) release, these plants can be compared to fossil fuel fired plants that would retrieve only one percent of the caloric content of the fuel (say coal) and turn the rest into waste. With such plants, coal resources would have been exhausted a long time ago.

Clearly, this kind of nuclear power plant is not the way forward if nuclear energy is to be of any use in the long term.

Fortunately, there are solutions to improve on them. The non-fissile U-238 that constitutes the major part of Uranium ore can be turned into fissile Plutonium, Pu-239, by neutron irradiation. This process actually takes place in all nuclear reactors. It is possible to build so-called "breeder reactors" where the rate of fissile Pu-239 production is higher than that at which U-235 is being burned. Such reactors produce energy and at the same time more fissile material than they consume. Eventually, about 75% of the original U-235 is used effectively. Instead of lasting for less than a century, Uranium ore resources can then last for thousands of years, and at the same time release a smaller amount of radio-active (and shorter life-time) waste.

At this point, it would seem that breeder reactors (or rather fast breeder reactors because they use more enriched Uranium and fast neutrons) are the only serious option for a long term electricity production by fission nuclear power plants. Some experimental reactors of that kind have been built (such as Super Phoenix in France), but their development in western countries has been stopped because of environmental concerns. They require extensive fuel treatment to recover Pu-239, itself the prime material for nuclear arms. Large scale development of fast breeder reactors and associated chemical treatment plants could only be done under close international supervision and control, whose mechanism remains to be established. More research and international collaboration is needed if this interesting and promising option is to become a reality.

A point that has been missed by both proponents and opponents of nuclear energy is that from the standpoint of the amount of entropy released per unit of energy generated, there is no fundamental difference between a conventional coal burning power plant and a nuclear one. The amount of entropy released is, in both cases, determined uniquely by the irreversible processes that take place during electricity generation, as we have learned from the analysis of the steam engine performed by Carnot. If the amount of entropy released is the primary concern, as may be the case, the technique used to produce the steam is almost irrelevant.

On the other hand emission of greenhouse gases is undeniably a serious problem. In the end, since we do need some kind of base load plant that produces electricity 24 hours a day all year, given the choice we might in the long term prefer nuclear power plants if the fast breeder reactor route can be developed safely. At least, I would.

6.4.3. *Transport of electrical power*

This chapter would be incomplete without considering the irreversible losses incurred during the transport of electrical power. These losses are equal to the electrical resistance of the line multiplied by the square of the current that it carries. There are two ways to reduce these losses: either by reducing the resistance of the line, which is possible if it is made of a superconductor; or by reducing the current, which one can achieve by increasing the voltage V on the line since the current I is inversely proportional to the voltage for a given transmitted power VI. Tesla pointed out in a breakthrough lecture delivered in 1888 that the voltage of alternative current could be increased by using transformers, and that this would make the transport of electricity economical over long distances, with acceptable losses. He had eventually the upper hand over Edison, whose low voltage direct current devices allowed only the transport of electricity over short distances.

Even with modern high voltage lines (700 kV lines are not uncommon), losses on the transport and distribution grids are not negligible. They vary a lot from country to country and from grid to grid, but as an order of magnitude one can retain a figure of the order of 10%. In the USA and the EU they are of about 7%.

The use of superconducting cables, whose electrical resistance is zero for direct current and very small for alternative current, may become interesting in the future for the reduction of losses over very long distances. Actually high voltage direct current lines are sometimes used for electricity transport over several thousand miles, with conversion from and to alternative current at both ends.

Chapter 7

Towards a World without Fossil Fuels

We have a kind of love-hate relationship with fossil fuels in general, and more particularly with oil. We cannot live without them — or at least we think so — but on the other hand we have become aware that their use is detrimental to our environment, and may even become life-threatening. We understand that the more we use them, the more damage we create. To make matters worse, we are also afraid of running out of them.

7.1. Increasing entropy and increasing energy needs

A few examples will illustrate this point. Carbon dioxide emissions and concentration in the atmosphere have increased due to the increasing use of fossil fuels. Higher CO_2 concentrations are thought to be responsible for global warming. To mitigate the effects of global warming we must use more air-conditioning. This requires a larger use of electricity, therefore of fossil fuels. And so on.

Pollution, or more generally entropy release, comes in two forms: heat pollution (release of low grade heat), and chemical pollution. Because we are more sensitive to the second one, we tend to take measures that decrease chemical pollution and forget that they will necessarily increase thermal pollution. We install filters at fossil fuels fired power plants, which lower their conversion efficiency, with the consequence that we need to burn more fuel to obtain the same amount of energy.

We prefer to burn cleaner fossil fuels — oil and natural gas — rather than dirty ones, coal. By doing so, we are depleting oil and gas reserves more quickly. Their price has severely increased, and soon we may not be able to afford them anymore. And now, what should we do?

All of the above are examples of the role played by entropy release.

Once this is understood, one also realizes that it makes no sense at all to talk of an energy crisis: since energy is conserved, there can clearly be no energy crisis. The term energy crisis is misleading as it gives the wrong idea of what is really going on, which is that entropy is generated by the many irreversible processes that accompany our activities. The increasing entropy increases our energy needs, which increase entropy release and so on.

7.2. The retreat of oil

There is no question that the massive economic development that has occurred since World War II was in great part made possible by the availability of abundant and cheap oil. From the mid-fifties onward, oil and its derivatives became the engine of modern society. Oil replaced coal in electricity generating power plants and for domestic heating. It became so cheap that nuclear electricity did not spread as had been expected, except in a few countries such as France and Belgium. The car industry took gigantic proportions as automobiles became the symbol of freedom for every one, rich or modest as the case may be. Instead of being reserved for the elite, air travel also became available to all. This is the world which we grew up in: it could not have developed without cheap oil. Are we going to have to abandon all these conquests because physicists say that burning oil generates entropy?

Entirely by accident, it turns out that exactly at the time when damage to the environment due to the use of fossil fuels (the common language used for increased entropy) is becoming a wide spread concern, questions are being raised as to the continuing availability of oil, or at least of cheap oil. Because of increasing costs, coal is back in favor. It is foreseen that the share of coal in electricity production will increase in the coming decades. In China where a new power plant opens up every week, not a single one is fired with oil. Are we witnessing the beginning of the retreat of oil? Maybe.

7.3. How much oil is left anyhow?

A lot of attention has been focused on questions such as: for how long will oil be available, when will world oil production reach its peak, how will the price of oil evolve and so on. So far, judging from past experience, experts in these questions have in general been unable to predict correctly the evolution of the market price of oil. Figure 7.1 shows three possible scenarios proposed by the US Energy Information Agency in 2006, from low oil price to high oil price. This figure shows on the left hand side the price history up to 2006, and on the right hand side projections up to the year 2030. In the high oil price scenario, oil price was projected to reach \$100/barrel in 2030. Actually, this price was already reached in 2008. Will it go back down to the reference or to the high price

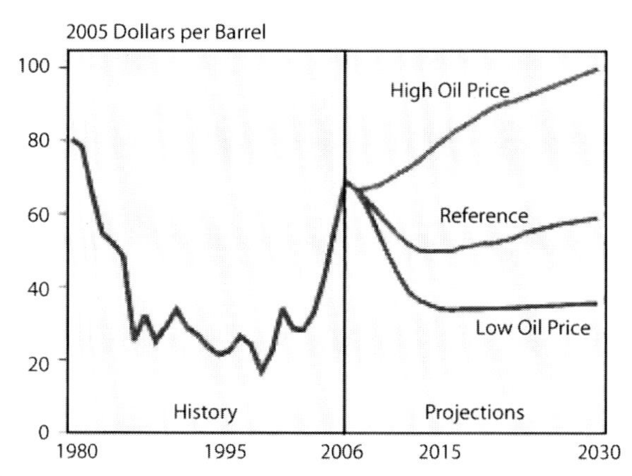

Sources: **History:** Energy Information Administration (EIA), *International Energy Annual 2004* (May-July 2006), web site www.eia.doe.gov/iea. **Projections:** EIA, *Annual Energy Outlook 2007*, DOE/IEA-0383(2007) (Washington, DC, February 2007)

Fig. 7.1. Past and predicted ranges of oil prices (per barrel). Note that in the high oil price hypothesis the price of a barrel of oil was not supposed to reach the US\$100 mark before the year 2030.

scenario, or will it keep increasing quickly beyond the $100/barrel mark, or will it remain more or less at its present level from $90 to $110 for a long time? I doubt that any reliable prediction can now be made.

One of the popular theories today is that of peak oil, called the Hubbert peak oil theory. It states that when production reaches its peak, half of the reserves have been consumed. The rate of increase of oil production has certainly slowed down since the 80's as can be seen Fig. 5.8. Some experts think that production is right now reaching its peak. The thought that half of oil reserves is gone, and that from now on production will start going down, as shown in Fig. 7.2, may be playing a role in the fast current price increase.

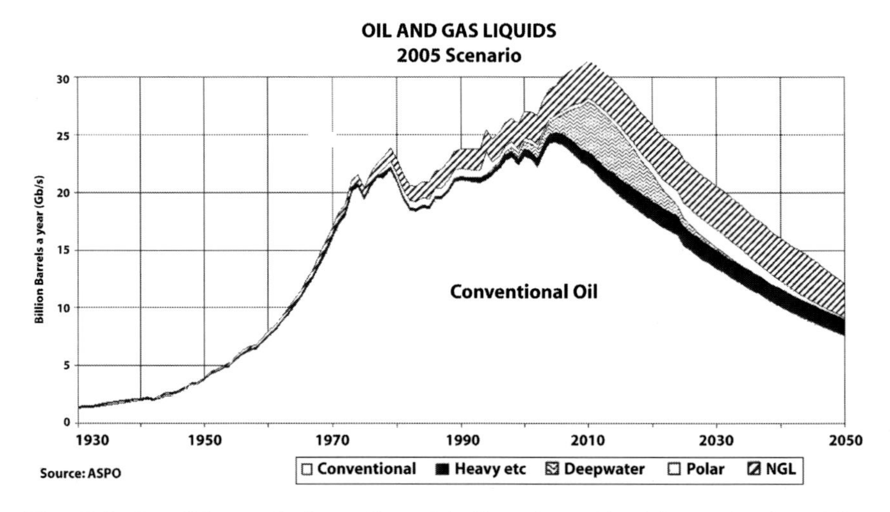

Fig. 7.2. Possible evolution of world oil and gas liquids production. The period of high oil production ranges roughly from 1950 to 2050. Beyond that, oil plays a minor role in the energy market (from Alex Kuhlman, Oil Peak).

But should oil remain roughly as expensive as it is to day, the exact amount left in the ground would become in a sense irrelevant. Oil has pretty much priced itself out of most of the energy market, whether it is because a large fraction of oil reserves are already depleted or any other reason. Coal energy is now three to four times less expensive than oil and gas energy. Coal reserves are much larger, production can be increased sufficiently to displace oil from most applications, certainly for electricity production and residential and office heating, may be even for liquid fuel production. The displacement will take time because huge investments are necessary, and meanwhile oil producing countries can charge any price they want. But the price difference is now so large that this displacement appears unavoidable. Whether heavier reliance on coal is a good thing is a different question.

It is interesting to note that coal reserves are much more widely spread world wide than oil is. The USA and China who are likely to be the world leading economies in the coming decades are coal rich and therefore will surely promote its use in a massive way. The USA have a tradition of relying on their natural resources without too much consideration for the environment. As for China, it is only now entering the modern economic era and has no choice but to rely massively on coal in order to keep its momentum. Forces at work in these two leading world powers are so powerful that there is little chance to stop this evolution, which is both technically possible and cost effective

Table 7.1 gives some quantitative information on coal reserves in the USA, Russia and China. Reserves in the three countries are substantial. China is leading in production and in fact would be the first to run out. Other important reserves are in India (84,000 million tons), Australia (82,000 million tons) and South Africa (49,000).

Table 7.1. Coal reserves are in millions of tons and production rates in millions of tons per year. Years of production left are for the current production rate.

Country	Reserves	Production rate	Years
USA	250,000	1,100	225
Russia	157,000	309	500
China	114,000	2,168	52

7.4. Replacing oil and gas by coal for residential heating?

Besides electricity generation, heating and transportation are the other two major areas where oil (or natural gas for heating) has so far remained the principal primary energy source. While cars have been powered by petrol since the invention of the internal combustion engine, oil did not become, in the West, the primary energy source for space heating until after World War II. Elderly people remember that homes were essentially heated by burning coal up to World War II and, at least in Europe, even later. In the end oil and gas won that market because they were cheaper, cleaner and easier to use. Nowadays, no building is heated by coal anymore in the western world.

But it may be only a question of time. Private homes and buildings could again be heated with coal, for exactly the same reason that new power plants use coal rather than oil to generate electricity: coal is now much cheaper for the same caloric content. Burning oil or gas to heat homes is quickly becoming so expensive that many people may not be able to afford it anymore. Cutting their heating bill by a factor of three would be highly appreciated.

Another heating option is to use electricity, but this does not make much sense from a thermodynamic point of view if electricity is itself produced by burning coal. In the process of generating electricity, about 60% of the calories content of coal end up as entropy rejected in the environment. Another 7 to 10% is lost in the transportation and distribution process of electricity. Contrary to the case of the motorcar, where the use of electricity is more advantageous, here thermodynamics are on the side of direct use of fossil fuel.

Coal is definitely less convenient and dirtier than oil, gas or electricity for space heating, but if it is much cheaper, one may wonder how long will we resist? Nowadays it has become fashionable to burn wood. Why not coal? Replacing our oil burning furnaces by coal burning ones does not require any major technical innovation. Filters could be installed to reduce pollution. Substantial savings in the amount of coal used are possible by using better construction standards as we have discussed in the preceding chapter. We may thus eventually see a decline of the use of oil and gas for residential heating, as these fuels become extraordinarily expensive.

7.5. Can we replace oil for transportation?

Replacing oil for transportation is a far more difficult issue. This is the sector where the use of oil, in the form of gasoline or diesel fuel, has been continuously expanding. For private cars, no alternative is yet available commercially. All flying airplanes burn kerosene, an oil derivative. If oil were to become unavailable, or extraordinarily expensive, our society largely based on a high degree of mobility would seemingly fall apart.

The all-electric motorcar, which we have discussed at some length in Chapter 6, offers an alternative that could fulfill part of our transportation needs with a much reduced entropy release. Besides, oil is now so expensive that gasoline can be produced at a lower cost from coal, using a chemical process that was developed in Germany in the thirties, and used by the Nazi regime when it got cut off from oil fields during World War II. The Luftwaffe did not stop flying because of a lack of fuel. China, which is rich in coal but poor in oil resources, is already building plants that will manufacture liquid fuel from coal. Apparently, this becomes profitable once the cost of an oil barrel becomes higher than US$30 or thereabout. If oil stays as expensive as it is today, it may be progressively displaced by coal even for transportation.

From our entropy standpoint, it makes of course little difference if we burn a fuel derived from oil, or from coal, whether it is for electricity generation, heating or transportation. Yet, it is somehow reassuring that our society will not collapse without oil. Maybe as an intermediate step it is in fact unavoidable that we shall turn back to coal or other hydrocarbons such as oil shale that are also less convenient than oil, while friendlier technologies are being developed.

7.6. Can coal be displaced as the major primary fuel?

It seems likely that in the next half century coal will displace oil and gas from most of their applications, may be even for transportation as discussed in the previous section. The resulting impact on the environment of a return to coal will certainly be negative as its combustion generates the same amount of entropy as burning oil or gas, roughly the same amount of CO_2, plus additional particles and gases. Under these conditions it is difficult to see how the amount of greenhouse gases emissions could be reduced, since total energy needs will evidently increase. In short, in this worst case scenario, coal would displace oil and gas fuels and in addition would provide most of the additional energy needed.

One can only hope that this will just be an intermediate stage, before total fossil fuel use can eventually be reduced down to an acceptable level. In the preceding chapter we have reviewed the various means at our disposal for achieving this aim. We shall now try to evaluate what could be their respective quantitative impact. In this chapter we limit ourselves to renewables, but will return to the role that nuclear energy might play in the following chapter.

7.7. Displacing coal with renewables

Large renewable energy sources — solar and wind for the most part — are available. As we shall see, they are more than sufficient to provide us with all our needs. The problem is that the energy that they provide us with is more expensive than that of fossil fuels as they come out of the ground.

We must get reconciled with the fact that the "production" of fossil fuels that nature has provided us with is necessarily cheaper than anything else. No energy source can ever be as cheap as oil or gas coming out of the ground by itself, no matter how clever we are in developing new technologies.

For many years, oil was so cheap that there was no incentive to develop alternatives to it. Even the development of nuclear energy was basically stopped because oil became cheaper, well before nuclear energy was banned for fear of radioactive pollution. Fortunately, there is now an enormous gap between the cost of production of oil and its market price. This gap is more than a factor of 10. Nuclear energy is now cheaper than the market price of oil and gas, some renewable energy sources are getting close to it. This is a dramatic change compared to what the situation was only a few years ago. Coal is and will probably stay cheaper than renewable energies for quite some time, simply because the large available reserves reduce the gap between production cost and market price. But in the long term one can expect that its price will also increase considerably. In the mean time, it may provide us with a manageable transition period from an economic standpoint.

7.7.1. *Competition for land space*

Except for geothermal sources, the development of all other renewables implies the massive use of land space. This is true for biomass for heating, biofuels for transportation, for wind and solar energies.

Agriculture, forests and renewables will inevitably compete for land space, and are in fact already competing. We believe that it is this competition and its consequences that will eventually determine how much we can expect from renewables. On-land wind farms are encountering increasing resistance from local communities.

The actual impact of renewables on our entropy balance must be evaluated taking into account their global effect, and not only the local one. In order to avoid detrimental indirect effects resulting from competition for land space, renewables should preferably be implemented on land areas that have no other use.

7.7.2. *Production potential of solar power in desert areas*

Deserts in sub-tropical regions, such as the Sahara, the Arabic peninsula, the Sinai, Southern California and more, offer the greatest potential for the production of renewable electricity and hydrogen. Production of solar electricity does not require water and in fact is most effective in places where there are few clouds and rainfall because this enables the use of mirrors for light concentration, as we have seen in the preceding chapter. Low production cost for electricity can then be achieved, without any competition for land.

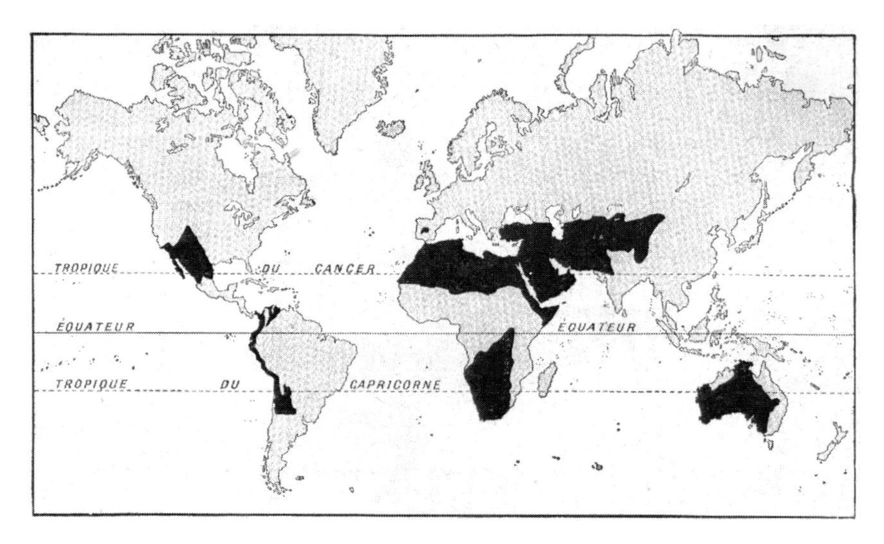

Fig. 7.3. Desert areas favorable for the installation of concentrated thermal solar electricity power plants, already identified in 1923 (in Le Ciel, Eds. Larousse, Paris).

Figure 7.3 shows in dark the areas where the implementation of solar power would be most efficient. There are good news and bad news in this figure (which, incidentally, was drawn almost one century ago). The good news is that there is plenty of space available for solar power. The bad news is that favorable regions are for the most part far away from densely populated areas where power is needed.

Installation of solar fields in desert areas will inevitably have some undesirable consequences. Even there, as we have learned from Saint-Exupery, there is life and it will be affected by their presence. Sand has a high albedo (it reflects into space a large fraction of the incoming radiation); replacing it by collectors over large areas will modify the energy balance. Collecting energy and dissipating it eventually, instead of reflecting it, will increase entropy release. Large scale collection of solar radiation may also have an impact on the local climate. Surely all these aspects deserve attention before any large scale solar plants will be constructed in deserts.

But let us now evaluate their potential for electricity production. The average solar radiation power falling per year on a square meter in the Sahara is somewhat higher than 200 W (it reaches up to 280 W in its hottest spot in the Niger). This means that the radiation energy falling on 10 square meters in the Sahara is equal to the average worldwide power consumption per person, which is 2 kW as we have seen in Chapter 3. Solar thermal power plants have now a conversion efficiency of about 30%. Hence, one would need about 30 sqm of desert land to cover the energy needs of one person. If one were to produce in the Sahara all the power needed for all humans, one would need to cover about 200 billion sqm with solar collectors. This area is huge, but still small in comparison with the total area of the Sahara, which is 9000 billion square meters. We need only a few percent of the Sahara's surface to provide all energy needs to all of humanity. Such a small fraction would presumably not even have a strong ecological impact. And surely we will not need to produce all of our energy in the Sahara. There are also hot deserts elsewhere. For instance the Mojave desert that covers parts of Southern California and neighboring states has an area of 57 billion square meters where solar thermal power could provide a substantial fraction of the energy consumption in Western USA.

These numbers are rather reassuring. In addition, as already said the energy pay back time for solar thermal plants is only 5 months, which is highly favorable. The main problem that remains to be solved is that of economical transportation of the power produced over very long distances. A little more on this in the next chapter.

7.7.3. *The potential for wind power production*

Wind power is at the moment the most widespread renewable. From an entropy point of view it is an ideal one: wind would in any case lose its power somewhere, and entropy would be produced, as we have noted in the previous chapter.

The main challenge is to extend the penetration of wind power to cover a large fraction of the total power needed. Considerable wind power resources are available, particularly off shore where they do not compete for space, but wind's variability and limited predictability make it difficult to exploit them in the absence of storage capabilities. As long as wind power is limited to a few percent of total installed power, the grid itself can "absorb" the effects of variability. At penetration levels of the order of 10%, some penalties already occur. Little is known about penetration levels of 20% and more. The consensus seems to be that 30% constitutes more or less a practical limit of penetration of wind power.

The penalty one has to pay at high penetration levels comes about in different ways. Since power is needed in the grid at all times at levels set by demand, conventional power plants must be added to the grid to take over in case of reduced wind power. Vice versa, wind power must be turned down when production exceeds demand. Additional power lines must be introduced in the grid to export electricity produced by wind turbines, sometimes over long distances.

Figure 7.4 gives an idea of wind's variability, compared to load, in Western Denmark.

Capacity credit (the saving of conventional installed power) is of the order of 40% of wind installed power at small penetration, and decreases as penetration increases. Figure 7.5 shows results of a study conducted in Ireland. At 500 MW installed wind power the capacity credit is close to 200 MW or 40% of installed power. At 3,500 MW installed wind power, the capacity credit is about 500 MW, or only 16% of installed power. There is clearly a law of diminishing returns: beyond a certain level it does not pay to install additional wind power, the capacity credit saturates. Typical numbers for capacity credit are in the range of 30% to 40% at low penetration, going down to 10% to 15% at 20% penetration.

Wind power, just as solar power, is in principle available in amounts that can easily exceed needs. It is its variability and

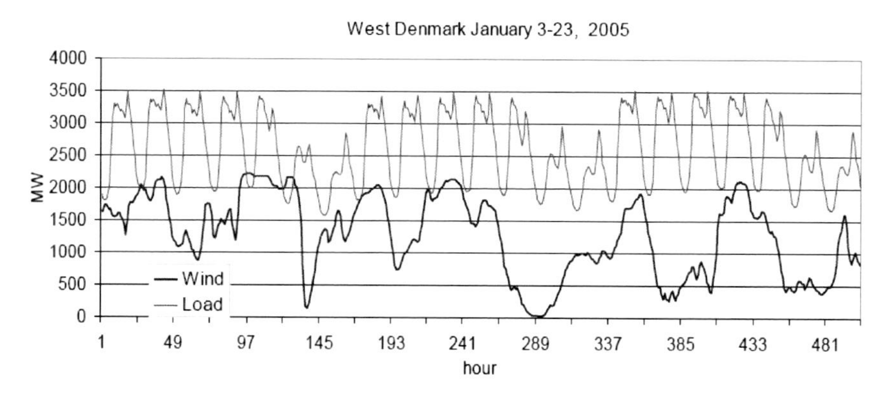

Fig. 7.4. Installed wind power in Western Denmark is 2,400 MW. Average power produced is substantially less. Load varies from 2,000 MW to 3,500 MW. Available wind power sometimes approaches zero for periods of up to a day. Power must then be available on the grid from other sources.

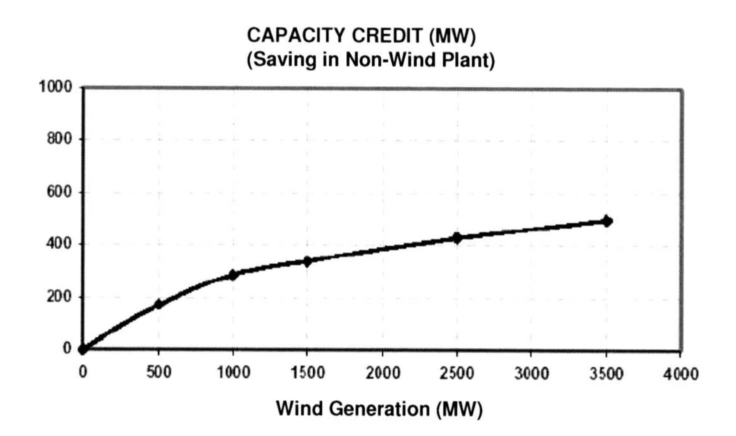

Fig. 7.5. Capacity credit of installed wind power from a study conducted in Ireland. The capacity credit saturates at high installed power.

limited predictability that limit its actual penetration in terms of actual energy produced.

Figure 7.6 shows the results of several studies on the variation of capacity credit as a function of wind power penetration. Numbers vary significantly from one study to another for large penetration values. At 30% penetration, capacity credit ranges from 20% to 5%.

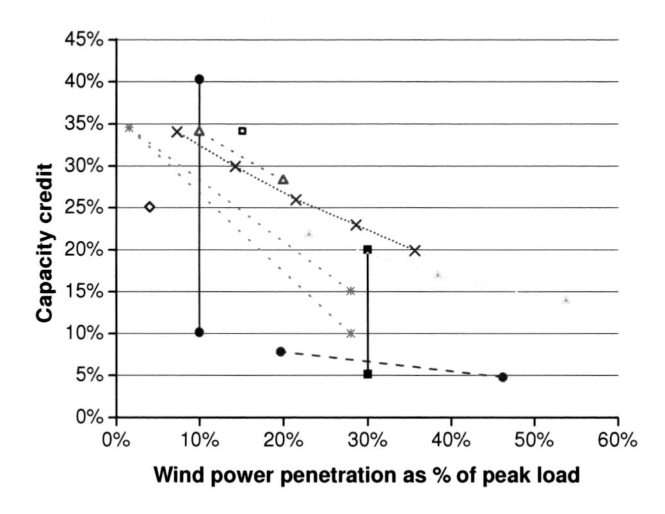

Fig. 7.6. Capacity credit for installed wind power as a function of penetration. At low penetration levels it is of about 35%, meaning that a 1 MW wind turbine will on the average deliver a power of 350 kW. The capacity credit decreases at high penetration levels, but with a large spread between studies conducted in 5 countries (Germany, Norway, Ireland, UK, US). After IEA Task 25 report "Design and operation of power systems with large amounts of wind power", 2007.

Germany has the largest installed wind power in the world as compared to its total electricity generating capacity. It is therefore instructive to look at the numbers in that country. The installed wind power in 2006 was slightly above 20,000 MW, while the total generating capacity is about 120,000 MW. This corresponds to a penetration ratio of about 17%. Electricity generated by the wind turbines was 6% of the total energy produced. Hence the capacity credit was about 30% of installed power, in line with numbers shown in Fig. 7.5. Germany has plans to further increase installed wind power. It will be interesting to see how the capacity credit will vary then.

7.7.4. *Distributed renewable power*

Solar water heaters and PV panels installed on roofs of residential buildings can make a substantial contribution to the total power available. This contribution is detailed in the next chapter. What we would like to point out here is the possible role played by legislation for a further extension of these distributed power sources.

7.7.4.1. *Solar water heaters*

Solar collectors for the production of domestic hot water have been in use for many decades in countries such as Japan and Israel. In Israel, their use has even become mandatory in apartment buildings since the mid-eighties for the deliverance of building permits. Roofs of all modern apartment buildings are now equipped with such collectors.

On average they provide about 80% of hot water needs. Typically, collectors having an area of a couple of square meters are sufficient for one family.

Insolation in Europe is on the average only 30% less than it is around the Mediterranean, the area of the collectors should just be increased accordingly. As explained in the previous chapter, techniques have been developed to deal with the problem of freezing water. They increase somewhat the cost of the installation but still in an acceptable way.

7.7.4.2. *Distributed photovoltaics*

Several countries, particularly in Europe, have adopted legislation that encourages home owners to install photovoltaic panels on their roof. The adopted scheme is such that the utilities will buy back the electricity produced at a rate that corresponds to cost, calculated on the basis of capital spent in the installation, the amount of electricity produced and current interest rate charged by banks. This electricity rate is at the moment about four times higher than that charged by the utilities. In other words, this scheme recognizes that distributed photovoltaic electricity is now about four times more expensive than that delivered by the utilities.

From the home owner's standpoint, distributed photovoltaics are in general less attractive than solar water heaters. For a given power, the investment in photovoltaics is 5 to 10 times higher than in solar heaters. The energy pay back time is also longer, being of the order of 2 to 4 years depending on the local average insolation, with current crystalline Silicium cells.

Nevertheless distributed photovoltaics have a couple of important advantages over solar water heaters. Increasing the area of solar heater collectors is not very useful: one could provide space heating in the winter, but one would not know what to do with the large amount of hot water produced in the summer. By contrast, electricity produced by photovoltaic panels covering up to a substantial fraction of the area of the roof is useful, as electricity can be exported to the grid. In other terms, one can make better use of the available roof area with photovoltaics than with solar water heater collectors.

Photovoltaics have another important advantage in areas where a substantial amount of the electricity produced in the summer is used for air-conditioning. Here photovoltaics would be competing with peak load rates, which can be several times higher than the average rate charged by the utilities. Even at its current cost, photovoltaic electricity may become fully competitive. Several kW of electrical power are necessary to cool large private homes. This power can be delivered by several 10 square meters of photovoltaic panels, an area that can be accommodated on the roof. Because utilities may in some areas face a serious problem to deliver the peak power needed at the hottest hours of summer days, and may even sometime have to cut power to avoid blackouts, it would be a relief to them if a fraction of that power would be produced by home owners. One could even think of legislation that would make building permits for large private homes conditional on the installation of a certain amount of photovoltaic power.

7.7.5. *About costs*

As we have shown, there are plenty of renewable energy sources on this planet, mostly solar and wind, largely sufficient to cover all our needs. Intermittency problems can be solved by coal or nuclear energy power plants that can provide base load power, at a level safe for the environment. The problem is not the availability, but the cost of renewable energy sources.

At this stage, they are about competitive with oil and gas, but not with coal and nuclear energy, with the exception of solar water heaters. Their large scale implementation requires at the moment some form of subsidy.

We must however accept the idea that in the future energy will necessarily be more expensive than it was during the age of cheap oil. In fact, it already is. How much more shall we have to pay? We can already give an upper bound estimate, based on the current cost of photovoltaics: photovoltaic electricity is today about 4 times more expensive than electricity delivered by utilities, generated from a mix of coal, gas and nuclear electricity. Solar electricity from concentrated solar thermal plants is maybe twice as expensive, and wind electricity about 50% more expensive.

These are very rough numbers. They will change with time as progress is being made.

It seems that the economy will have to adjust itself to an electricity cost about twice to three times higher than what we pay currently. Since, as the oil age winds down, more and more of the energy that we use will be delivered to us in the form of electricity, its cost will be the determining one for the energy market.

Chapter 8

A Changing World

Life is at the center of the so-called energy crisis. The golden rule for the preservation of life is: do not dump into the environment more entropy that can be taken care of by the available energy input. The more entropy is dumped, the more energy is needed to restore order. This rule holds at all levels, from the household to the entire biosphere.

At the level of our household, the energy supply is provided to us by society in the form of food, fuel, and electricity. At the other end, society dumps its entropy in the biosphere. At that level, the energy input from solar radiation is the only source that can restore order. There is of course a limit to what it can do. If that limit is crossed, disorder will increase in the biosphere. Climate stability will be threatened. Life eventually might be threatened as well.

In primitive societies, man dumped entropy directly in the environment, in the form of polluted water and soil, carbon dioxide rejected into the air, and debris of the modest artifacts that he used. Nature — what we call today the biosphere including the energy supply from the sun — easily took care of it. In some parts of the world, populations still live like that: the energy they consume per capita is very small, less than the food energy they consume.

By contrast, in modern societies there are a number of levels at which entropy is released — the household; the working place; the city and its networks for water and electricity supply; private and public transportation; used water treatment plants; garbage collection and disposal; the state or country with its power plants, electricity transmission and distribution, long distance transportation networks; international trade networks at the scale of the planet. Entropy is released and energy is supplied at each level and the golden rule should apply at all scales. But in the end, all the entropy is released into the biosphere, and again only energy supplied by solar radiation can restore order.

Man in primitive societies was in direct contact with the biosphere and could experience the results of his actions. The complexity of modern society and its many organizational levels shield us from this direct contact and experience. One can wonder if climate change is a serious concern for many. Should temperature rise, one could always add more air-conditioning; should it go down, more space heating. These protective actions will only make the global situation worse but, at the individual level, this is not felt. In practice, the only inconvenience is the increase of the cost of energy, including of course food energy. Nobody likes prices to increase but, from the standpoint of thermodynamics, the increase of the price of energy is the best news we have had recently. We give below a few examples of how it can lead eventually to substantial reductions in entropy release.

What we would like to do in this concluding chapter is to give some perspective, suggest what a realistic objective could be, then show on the one hand the opportunities, and on the other, the dangers we face. Some of the foreseen changes will possibly occur for economical reasons — but after all there may be a link between physics and economy.

8.1. A realistic objective

As we have seen in Chapter 3, there is a vast disparity between power consumption per capita in different parts of the world. While the average is equal to 2 kW, it varies from more than 10 kW to 100 W.

Based on the main conclusion of Chapter 5 on Climate Change, a reasonable objective would be to keep the average power consumption at the level of 2 kW, with half of it coming from renewables. This would reduce greenhouse gas emissions to their level of 1970, a level that we have judged acceptable since up to 1970 anthropogenic effects on the climate where hardly noticeable according to the last IPCC report. We therefore believe that this level might be adequate to ensure long term climate stability.

As we discuss below, we believe that from the supply side this level of renewable energy — 1 kW per capita — can be achieved without too many difficulties.

But from the demand side, it would mean for developed countries a progressive reduction of the energy consumption down to 2 kW. We will summarize some of the directions that can lead to a substantial reduction of the demand level in developed countries, which we have discussed in Chapters 6 and 7. But the needed reduction will likely prove to them to be very difficult to achieve, because it probably implies modifications of their organization at the society level.

Of course differences in local climate should be taken into account: it is clear that the energy consumption will have to stay higher in very cold climates than in warmer ones, but we may take the 2 kW figure as a guideline.

8.2. The supply side

Let us now consider how best to supply the 2 kW power per capita. Again, we emphasize that this is the power consumed per capita at the society level, not at the household level.

8.2.1. *Distributed power supply*

Distributed power supply installed on the roofs of buildings and private homes can play a substantial role.

Let us summarize what contribution we could get from such installations.

Solar water heaters can provide most of the energy needed to heat up sanitary water. This represents a contribution of 100 W per capita, or 5% of our allowed 2 kW. The roof area needed is of the order of a few square meters for a family of four, which allows the installation of solar water heaters on the roofs of most residential buildings. Legislation making solar water heaters mandatory on apartment buildings of less than 10 stories can be very effective to spread this technology, as shown by the example of legislation passed in Israel in 1985. Similar legislation has been passed in Spain recently.

Photovoltaic panels of larger areas can be installed on the roofs of single family homes, or two to three story apartment buildings and public buildings. An area of 60 square meters which can be installed on the roof of a typical single family home will deliver about 10 kW at peak power, or around 50 kWh per day, which corresponds to an average power of 2 kW, with the exact value depending on local insolation. In a single home family of four, this corresponds to 500 W per person, or 25% of the 2 kW goal. This is more than the electricity needed at home. Residential buildings will then become net energy exporters.

Taken together, solar water heaters and PV panels installed on roofs could cover up to 30% of the total energy supply of families living in single family homes. They could cover more than 10% of the needs of the total population.

8.2.2. *Wind power*

Large scale wind farms are visible signs of a changing world. The introduction of wind power has been truly impressive over the last decade. Tens of thousands of megawatts have been installed, more than nuclear power plants. More will be installed in the future in many countries all over the world, in Europe, in the United States, in China and so on.

The main advantage of wind is that it is widely available. Its main disadvantages are its variability and limited predictability. They reduce effectively the economically viable penetration of wind power. Drawing from the experience gained in Germany and Northern European countries, wind can cover about 10% of the total electricity demand (which is roughly achieved with 30% of installed power). The installed wind power in Germany is at the moment about 250 W, or 80 W produced per person. This is of the same order as what one can get from solar water heaters. It is quite realistic to assume that wind power will contribute of the order of 5% of the desired total power production.

8.2.3. *Large scale solar electricity*

There exists a huge potential for solar electricity generation in desert areas. In subtropical deserts such as the Sahara, the Lybian and Egyptian deserts, the Sinai, the Negev and the Arabic peninsula, solar radiation is available almost everyday of the year with great predictability. The absence of clouds allows to concentrate effectively solar radiation, which lowers electricity cost considerably whether one uses solar thermal or a photovoltaic concentration scheme.

Large scale solar thermal concentrated fields have been operated successively in Southern California for the last 25 years. Similar technology could be implemented in the deserts that extend over 5 time zones from Western Sahara to the Arabic peninsula. Solar fields could provide all the electricity needed by Africa and Europe without interruption during almost 12 hours a day. Because these hours correspond to peak demand, most of the total energy needed could be provided in this way. Only one percent of the deserts areas would need to be covered by the fields.

Altogether, the combination of distributed power supply, wind and concentrated solar power could provide up to 1 kW of power per person, or 50% of the wished total power consumption, with very little entropy production. The rest, which is within the allowed limit of 1 kW, will have to come from a combination of coal and nuclear electricity.

8.2.4. *The importance of improved electricity networks for the implementation of renewables on a large scale*

This is the place to remember that one of the advantages of oil is that it is easily transported and distributed. The advantage of an easy and cheap distribution network is not something to be relinquished lightly.

Moving toward a world where less fossil fuel will be burned requires extending the role of the electricity network. More electricity will be produced, transported and distributed, particularly in developing countries. A few decades from now, the electricity network may look quite different from what it is today.

As we have described in Chapter 6, the idea of an electricity network is due to Tesla. The concept of the network was, and still is, that a.c. electrical power is generated in central power plants, voltage is raised through transformers and power is then transported. At the other end, voltage is reduced through step down transformers and distributed to end users as low voltage a.c. current.

With the introduction of renewables, power generation is becoming more distributed in many different ways.

Some of it will be generated in residential buildings.

Large wind farms are being built in Germany tens of miles away from the shore in the Northern Sea, from where power needs to be transported before it can be injected into the network.

Solar thermal power plants require concentration and will therefore preferably be built in desert areas where the skies are clear most of the year — but deserts are very thinly populated and again the produced power will need to be transported away to where it is needed.

There will therefore be a need to implement changes in the electricity grid. Devices for handling variable power sources will need to be developed. More attention will have to be given to long distance, low cost electricity transport.

With conventional power lines, losses over distances of several thousand kilometers may be too high to allow renewables to be cost effective. As discussed in Chapter 6, high voltage direct current lines may then become preferable.

Superconducting cables may offer in principle a long term solution. Unfortunately the superconducting state is only reached at temperatures much below ambient. Up to a few decades ago, these temperatures were not much above absolute zero, which is –273.16 degrees Celsius below ambient. Recent discoveries have raised it to about –180 degrees below ambient. This is an enormous progress, but probably still insufficient to reduce the cooling energy requirements down to an acceptable level.

A futuristic combination could be that of superconducting cables to bring electrical power from deep into the desert up to the sea shore, and hydrogen production by electrolysis of sea water followed by hydrogen liquefaction and sea transport. There would be no net entropy produced.

8.2.5. *Switching back to coal*

The new, indisputable evidence that the oil age is winding down does not come from theories such as peak oil, or from new evaluations of proven reserves. It comes simply from the price of a barrel of oil, or more exactly from the price of a calorie of oil compared say to a calorie of coal. Oil was four times more expensive than coal as delivered to electricity generating plants in 2006, as can be seen Table 8.1. This ratio is even higher now, closer to 6, at the time of writing. Experts may be divided as to how much oil is left in the ground, but none will today take the risk of predicting that the price of oil will go down, relative to coal, to where it was a few years ago.

Table 8.1. Prices of energy for different fuels in US$/Million BTU.

	Oil	Gas	Coal (NYMEX)
2003	6	5	1.5
2004	8	6	2.5
2005	12	8	2.5
2006	15	7	2.5
2007	14	6	2.5
2008	28	13	6

Table 8.1 shows a comparison of the price of different fossil fuels measured (1 MBTU = 0.293 MWh). Up to the year 2004 the price of oil was not much higher than the price of natural gas, both were 2 to 3 times higher than the price of coal. In the last year, all prices have about doubled, that of gas being about half the price of oil, and the price of coal about one quarter.

It is our view that coal will and should now replace oil in most of its applications. It will because it is cheaper, and it should because oil is a much more precious material. We need it for making plastic materials for instance. In these processes oil is not degraded in the same way as when it is being burned. Long molecules making up plastic materials are still preserved, and in many instances can be recycled. In other terms, less entropy is generated in these applications than in combustion. We also need oil derivatives for making fertilizers.

Burning oil for space heating makes no sense. It is preferable to burn coal, in case temperatures are too low for heat pumps to be effective. The heating bill for many who use oil or gas for space heating is now so high that if a distribution network for coal still existed, as it did up to 50 years ago in many countries, many people might be tempted to switch back to it.

Coal has got a bad reputation, because it used to be at the origin of the London fog. But improvements in coal combustion make it now much more environmentally friendly than it used to be. For many years to come, coal will again be the primary energy source for electricity generation power plants. This cannot be avoided anymore. Nuclear power plants could compete with coal fired power plants, but their construction takes so long — from 15 to 20 years — that new plants will have no impact for the next 20 years at least.

Oil should be conserved.

8.2.6. *Nuclear energy as a replacement for coal fired plants*

After a strong start at the end of World War II, the construction of new nuclear power plants has been all but abandoned by most of the main industrial countries, with the noticeable exception of France. Low price of oil and gas and worries about safe long term disposal of nuclear waste may have been the main reasons for this change in attitude from the eighties onward. Renewed interest in nuclear electricity worldwide is evidently the result of higher and unpredictable oil and gas prices.

But in the mean time, precious time has been lost. Safe waste disposal requires more research. Thermodynamic efficiency of current nuclear power plants is rather low, about 30% (less than that of modern coal fired plants), more research and development is necessary to increase it. In view of the limited availability of natural Uranium, it is necessary to develop the fast breeder reactors that we have discussed in Chapter 6 if nuclear electricity is to remain an option for the long term.

New massive and international R&D programs are necessary to keep the nuclear option open. It should consist of two distinct programs.

In the first place, disposal of the vast amount of radio-active waste generated by the reactors built in the eighties must be tackled. At the moment this waste is kept as surface storage. At some point deep storage will have to be performed. It would make little sense to resume massive building of this kind of plants, because they would exhaust quickly the natural Uranium ore and at the same time produce more waste. A limited number of plants with improved efficiency and sub-unity but higher breeding efficiency could be built as an intermediate stage before fast breeders become available.

In parallel, more research should be conducted on fast breeder reactors with considerable international collaboration, since in the end breeders and the associated chemical treatment plants will only become acceptable if operated under strict international supervision.

For the foreseeable future, it is likely that many more coal fired than nuclear power plants will be built. But in the long term nuclear energy may be the main practical option for providing the 1 kW per person base load electricity generation that needs to be supplied within the general 2 kW framework.

8.3. Reducing the power consumed in developed countries

The objective of a 2 kW power consumption in developed countries will not be easy to achieve but it is not unrealistic.

Installed electrical power per capita is about 3 kW in the USA and 1.5 kW in Western Europe. It corresponds to one third of total power consumption. The other two thirds go mostly into space heating and cooling, and transportation. Improved building insulation and a shift away from the internal combustion engine could bring us close to the desired objective of 2 kW, which would be mostly delivered in the form of electricity. Both are in principle possible.

8.3.1. *The case for the electric car*

The most difficult obstacle towards fulfilling the 2 kW consumption objective may be to reduce considerably the power spent on transportation.

The electric car could be an important step in that direction. A quick calculation shows that the cost of a kWh of electricity is now much cheaper than that of a kWh of gasoline when it comes to car propulsion. The energy content of one liter of gasoline is 10 kWh, but we only get about 2 kWh of mechanical work out of it because of the low conversion efficiency of the internal combustion engine. The price of one liter of gasoline varies according to local taxation practices from less than US$1 to more than US$2. The half liter that gives us 1 kWh of work costs therefore from 50 cents to US$1, or more. On the other hand, a kWh of electricity costs around 10 to 20 cents in most countries This is roughly five times less than the kWh that the internal combustion motorcar gets out from gasoline.

The electric car always had the upper hand from the point of view of entropy release, for reasons that we have discussed in Chapter 6: it uses electricity coming from central plants where fuel is burned much more efficiently than in the internal combustion engine, and it can regenerate part of the electrical power that it consumes. But up to five years ago, the electric car had no economic advantage over the internal combustion engine motorcar. Now, for the first time, it is also economically feasible. This is one case where high oil prices could turn out to be highly beneficial for mankind. An all-electric car with ideal batteries, an ideal electrical engine and no friction losses whatsoever, would regenerate as much electricity as it consumes. Even if not ideal, it will be a much better thermodynamic machine than the internal combustion engine motorcar. As progress is being made on batteries, this should translate into an increasing economic advantage, as well as a diminishing power consumption and entropy release.

We have estimated that the energy consumption of an electric car can be as low as 1 kWh for 10 kms. For a car driven 15,000 kms per year this corresponds to an energy spent per day of 4 kWh, or to an average power consumption of about 200 W, or 50 W per capita if this car is the family car, to go back to the example given in Chapter 3. This compares to an energy spent per day of 40 kWh, or to an average power consumption per capita of 500 W for the conventional car, a huge saving.

Additionally, electric cars may provide the only way to block the growing use of biofuels and its disastrous effect on food supply. Massive introduction of electric cars would immediately solve this acute problem. It is true that in a first stage the use of electric cars would increase the use of coal. But we strongly believe that it is preferable to run cars on electricity provided by coal than on biofuels. Besides, at a later stage they could be run on renewables, or nuclear electricity.

Running a car on electricity produced by photovoltaics has in fact now become cheaper than running a car on petrol. In countries where legislation makes it mandatory for utilities to buy back photovoltaic electricity at cost, an ideal scheme has become feasible: put PV panels on one's roof, sell the electricity to the utilities at an attractive rate when it is produced, and buy back part of it at night at a lower rate to recharge the car's batteries. This is quite a smart combination since the grid is used in effect by the owner of the PV panels as a means to store the photovoltaic electricity it has produced, while allowing the Utility to get electricity from PV installations at the time of day where it needs it most. This combination will allow transportation with almost zero entropy release, an enormous improvement compared with the present situation.

Technically, the only obstacle still preventing a wide spread use of electric cars is their battery bank. Batteries are heavy and perhaps more seriously, have a limited life time when submitted to heavy charging-discharging cycles, as we have discussed in Chapter 6. But this disadvantage will be more than compensated for by the large savings in fuel costs.

This is one change that we will hopefully see in the not too distant future: PV panels on our roofs (or on the roofs of public buildings) to produce effectively electricity for our cars, by using the grid as a buffer.

8.3.2. *Family energy budget and power spent at the society level*

With the electric car, power spent per capita on transportation becomes of the same order as that spent on food, electrical appliances or hot water, or about 100 W per capita for each of these items. The dominant power budget line is now that for heating — even with a well insulated home it is of about 300 W per capita. With some improvement on the efficiency of electrical appliances, it may be possible to bring the total power spent per capita in the framework of the household down to 600 W. This looks good compared to the proposed 2 kW level, but it is not the whole story.

As we have already noted in previous chapters, power spent at the society level is several times higher than that spent at the level of the household. For instance, power spent in power plants is several times higher than the electrical power delivered to the household. If a power of 600 W is spent at the household level in the form of electricity, a power of about 1,800 W is spent to generate and transport it.

In addition, society provides services in the form of food production and distribution, public transportation, education, health care, water supply, garbage collection and treatment, administration of different kinds, that all need power. These services would need to be analyzed carefully from the standpoint of power spent and entropy released. They will need to undergo very significant changes if the 2 kW objective is to be met. For instance, food production and its transportation consume much more power than the food energy we actually consume. As said before, the "beef machine" is a very inefficient one. In western societies, our 100 W food energy intake may easily turn out to correspond to 1 kW of power spent. This is much more than the power we will need to run the hoped for electric car. It will have to be reduced, which implies major changes in the food industry and our dietary habits.

8.4. The dangers

A climate run away and food supply problems are amongst the most serious dangers we are now facing.

8.4.1. *Is there a climate run away?*

It would seem that at some point in time we have crossed a line and moved into unknown and dangerous territory where the total entropy being released into the biosphere is too large to be taken care of by solar radiation. Looking at simulations published in the last ICPP report as shown in Chapter 5, it seems that this occurred in the mid-1970s. At that time temperatures started to rise quickly beyond values calculated excluding anthropogenic contributions. Also at that time, levels of CO_2 in the atmosphere started to rise quickly, well beyond values typical of previous interglacial periods. Some visible manifestations of warming, such as the melting of ice in the northern pole region, have accelerated beyond the predictions of computer simulations, which if confirmed may indicate that the climate models used are no longer appropriate. All of this suggests that we may have entered into a regime of climate instability. It would seem prudent to limit entropy release to pre-1970 levels.

Several scenarios have been proposed and studied that could lead to a stabilization of the level of carbon dioxide in the atmosphere. The 2007 IPPC report, aimed at giving guidance to governments, lists various means of conserving energy (well, again, it is conserved anyhow...) and of developing less polluting energy sources.

Unfortunately, this is not what is happening. From 1980 to 2005, emissions measured in millions of Carbon equivalent metric tons have *increased* from 18,330 to 28,192. Half of this increase is due to the contributions of China (3,868) and India (877). North America contributed 1,576, Africa and South America about 500 each and Japan 300. Europe is the only continent whose emissions have remained stable. Carbon dioxide atmospheric concentration has risen over the same period of time from 335 part per million (ppm) to 380 ppm, or at the rate of 180 ppm per century.

Available data (see Fig. 5.11) suggests that the biosphere reacts to perturbations like a living organism. It is capable of adapting to changes and remains stable up to a certain level of perturbations — there is no evidence for anthropogenic temperature changes up to 1975 — but passed that point it becomes unstable. A continuing increase in the use of fossil fuels and CO_2 emissions may be a serious threat to climate stability.

8.4.2. *Is carbon dioxide atmospheric content a sufficient indicator?*

Measuring the well being of the planet by the CO_2 atmospheric emissions, as often done these days, is much too simplistic and can lead to unfounded and dangerous conclusions.

First of all, this choice suggests that the climate varies smoothly with CO_2 content. A little more CO_2 will result in a little warmer climate. As discussed above, this may not be true.

Second of all, retaining CO_2 emissions as the sole measure of the virtuous character of a given type of energy production may be dangerous.

According to this criterion, nuclear electricity is credited with the highest mark since nuclear power plants do not emit any CO_2. Yet, on top of the known radioactive waste, they do also release entropy in the form of low grade heat rejected into the environment, which should be taken into account in the general entropy balance.

Another example is that of biofuels. They offer in principle the possibility of producing liquid fuel whose combustion will not increase the CO_2 content in the atmosphere, since the primary material they are made from (corn, sugar or others) is produced by photosynthesis. In reality, transformation of the primary material into liquid fuel requires an energy input, which is provided by the combustion of non-bio fuels. There is therefore a net CO_2 release into the atmosphere linked to the production of biofuels. Even more seriously, production of many of these biofuels comes at the expense of food production.

8.4.3. *Food supply*

Massive production of biofuels looms as one of the biggest and most immediate dangers that mankind is facing. We may have to choose between feeding our cars or feeding ourselves. Or, more exactly, between feeding our cars and feeding others.

Since oil prices have skyrocketed, it has become profitable to transform crops into liquid fuel. This is the very real incentive for the production of biofuels, rather than lowering CO_2 emissions. If we stick to the kind of cars we are used to driving, it is not at all clear whether the trend towards a massive use of biofuels can be reversed. They make everybody happy: the car owner who pays less for fuel; and may even think (wrongly in our view) that he is doing something positive for mankind when he chooses to go "bio"; the car manufacturer who can keep making money selling the kind of car he knows how to produce without much further investment in R&D; the food grower, who sees the value of his product skyrocket together with that of oil; and the politician who can claim that the tax incentives he is spreading around are enhancing the energy security of his country. A further and important advantage of biofuels is that their production can be increased quickly with modest capital investment — unlike building advanced nuclear power plants or laying the foundations of a hydrogen economy.

On the global scale, the trend towards a massive use of biofuels might be one of the worst developments we have seen in recent years. Combined with the recent (and hopefully temporary) reduction in crop production due to unfavorable climatic conditions, it is generating a fast decrease of food supply and a corresponding increase in food prices. For instance, a third to a half of the corn produced in the USA is now used for ethanol production. Palm oil is now used for bio-diesel fuel production. The consequences are already felt. Shortages in cooking oil have been reported. Food rationing has reappeared in Egypt and Pakistan. Indonesia, Yemen, Mexico have been mentioned as countries where hunger triggered by high food prices has appeared. A quarter to a third of the population of India is reported to be undernourished.

Announcements to this effect have been made by Ms. Josette Sheeran, director of the UN World Food Program, as reported by the Swiss newspaper “Le Temps” issue dated February 26, 2008. According to her declarations, the Program may have to curtail food distribution, and even to limit it to “persons who really need it”. This strange declaration may in fact just reflect the fact that for many years there was a large food surplus that was channeled through the Program to places where people could actually feed themselves, but at a higher cost due to primitive agricultural practices. These days are apparently over; one can only hope that past beneficiaries still remember these practices.

Even in rich countries, higher food prices are being felt. Public support for biofuels may hopefully wane when it will be realized that there is a link between saving money by filling up with biofuels, and having to spend more money on food.

Problems caused by a massive production of biofuels are now better understood, and one can hope that further expansion will be delayed until its impact can be fully assessed.

8.4.4. *Renewables and water*

This quick overview would be incomplete without a mention of another major problem, which is the massive need for water of drinking quality. It is well known that there exists a severe water shortage affecting hundreds of millions of people around the world. Water rationing is effective in many places.

Lack of water affects directly people's health, and also food production. In some countries in the Middle-East, a large fraction of potable water is produced by sea water desalination. Desalination processes rely at the moment on the use of fossil fuels to provide the necessary energy, either by thermal methods involving the separation of salt through evaporation, or by mechanical methods (reversed osmosis) using electricity provided by generators.

In the long term, the use of fossil fuels for water desalination is very problematic as it will further impact the climate. Use of renewables is the only reasonable alternative. Since water needs are mostly concentrated in countries with a warm climate, solar energy is the most appropriate solution. It could, in fact, be an excellent market for solar thermal electricity.

It is known that large amounts of underground salty water exist in desert areas such as the Sahara. Desalination using solar electricity that may become available in these areas for reasons we have discussed above would completely change local living conditions. The dual availability of cheap energy and water could make them attractive places to live in.

Back to Africa?

Index